Physics Wave Concepts
For Marine Engineering Applications

REEDS INTRODUCTIONS

Essential Sensing and Telecommunications for Marine Engineering Applications

Physics Wave Concepts for Marine Engineering Applications

Physics Wave Concepts
For Marine Engineering Applications

Christopher Lavers

Sara-Kate Lavers

ADLARD COLES NAUTICAL

B L O O M S B U R Y

LONDON · OXFORD · NEW YORK · NEW DELHI · SYDNEY

Thomas Reed
An imprint of Bloomsbury Publishing Plc

50 Bedford Square
London
WC1B 3DP
UK

1385 Broadway
New York
NY 10018
USA

www.bloomsbury.com

REEDS, ADLARD COLES NAUTICAL and the Buoy logo are trademarks of Bloomsbury Publishing Plc

First published 2017

British Library Cataloguing-in-Publication Data
A catalogue record for this book is available from the British Library.

Library of Congress Cataloguing-in-Publication data has been applied for.

ISBN: PB: 978-1-4729-2215-1
ePDF: 978-1-4729-2217-5
ePub: 978-1-4729-2216-8

2 4 6 8 10 9 7 5 3 1

Typeset in Myriad Pro by Newgen Knowledge Works (P) Ltd, Chennai, India
Printed and bound in Great Britain by CPI Group (UK) Ltd, Croydon CR0 4YY

To find out more about our authors and books visit www.bloomsbury.com. Here you will find extracts, author interviews, details of forthcoming events and the option to sign up for our newsletters.

The seas have lifted up, LORD,
  the seas have lifted up their voice;
  the seas have lifted up their pounding waves.
Mightier than the thunder of the great waters,
  mightier than the breakers of the sea –
  the LORD on high is mighty.
Psalm 93:3–4

Thanks

Chris – I thank Prof. Alan Myers for his support, Prof. JR Sambles FRS for advice in all things physics (especially optics), and lastly my wife and family.

Sara-Kate – I thank my dad for letting me work with him on this book, God for giving me the ability to do it, and His numerous blessings over me and my family.

Chris and Sara-Kate both wish to thank Ms Kirsty Schaper for her support with this book.

CONTENTS

Preface

This book aims to introduce the basic theoretical concepts underpinning the Electromagnetic Wave Motion topics covered in most maritime-related courses, whether Naval, Coastguard or Merchant Marine Engineering. It is also designed as a suitable mathematical foundation for key physics concepts, providing the necessary learning skills and knowledge prior to the study of Reeds Marine Engineering and Technology Series, Volume 6: *Basic Electrotechnology for Marine Engineers* and Volume 7: *Advanced Electrotechnology for Marine Engineers*. It is advisable that these and other maritime users have a basic understanding of the theoretical concepts upon which many essential modern sea-going sensors and telecommunications devices now operate.

This book covers the background material required in the Electrotechnology syllabi of the UK Department for Transport Examinations and is suitable as an introduction for Marine Engineering Cadets studying the Electrical Engineering Principles unit of the Business and Technician Education Council (BTEC) programme and other related subjects, as well as for those pursuing Further Education (FE) and Higher Education (HE) studies. This volume also provides a complete basic wave principles foundation for the Reeds Introductions volume *Essential Sensing and Telecommunications*. Additional basic wave motion questions to practise those in this book can be found in other volumes by the authors [P1–P2].

An understanding of basic wave motion concepts will be necessary to your future career and the book is designed to support your learning of topics covered within Physics Wave Concepts principles and also to help you develop your understanding of maritime technology related matters. This book targets electromagnetic wave applications but will also consider sonar acoustic and surface water waves where appropriate, showing the scope of the theoretical wave approach. The book will investigate key wave motion topics vital to International Maritime Organization (IMO) educational recommendations, which support your safety at sea.

Knowledge regarding electromagnetic waves and electromagnetic devices is an essential requirement for merchant navy sea service, particularly for the Standards of Training, Certification and Watchkeeping for Seafarers (STCW) qualification in Marine Electrotechnology (as Chief Engineer and Second Engineer), as mandated by the UK Department for Transport Maritime Coastguard Agency (MCA).

This introductory book has been written as simply as possible to support the large number of overseas students for whom English is not their first language.

References in square brackets throughout the text link to the references found at the end of each chapter.

REFERENCES

[P.1] *50 Basic Wave Motion Questions for Marine Engineers*, CR Lavers and Sara-Kate Lavers (Lulu.com, 2016, ISBN: 978–1-3263–5569–2).

[P.2] *50 Advanced Wave Motion Questions for Marine Engineers*, CR Lavers (Lulu.com, 2017).

For Teaching and Learning questions associated with Marine Engineering and Physics, Dr Chris Lavers may be contacted at: christopher.lavers@plymouth.ac.uk. He is particularly interested in individual case studies or vocational examples and any solutions *you* may have found to address maritime problems you have encountered.

Introduction

Have you ever sat on a beach on a warm sunny day, watching as the waves reached the shore, and wondered what is physically happening (figure I.1)? We see the water waves because they reflect visible light arriving from the sun, which enters our eyes. Simultaneously, you may hear the vibrations of the air molecules travelling into your ears as sound waves (transferred from a sound wave to a mechanical wave and then back to a biological acoustic wave in your head) before producing an electrical wave that the human brain interprets. Or perhaps you have listened underwater to the grinding acoustic waves produced as sand particles rub violently one against another, or felt vibrations set up in the sand with your hand as the waves strike the shore, travelling inland as weak seismic waves? During a tsunami, these normal wave processes may be amplified considerably but are essentially the same, although several different wave propagation methods may be at work at the same time.

Figure I.1: *Cape Comorin ('Land's End'), Southern India, looking south towards Antarctica. The beach was severely affected by the Boxing Day tsunami in 2004.*

As we observe the 'single event' of a wave striking the shore, it can produce a variety of different outcomes, or we might say that different energy exchange and energy loss mechanisms have been simultaneously activated. It is in this sense that we can describe water waves, light (and other electromagnetic waves), sound waves, acoustic and seismic waves in terms of various wave properties, from their size (amplitude) to their ability to be reflected, diffracted and refracted. Each type of wave mentioned here, and others, has a different impact on our human senses, and there are in fact many electromagnetic waves that exist far beyond the limits of our unaided human senses to detect, such as invisible X-rays or gamma-waves. In some cases the medium that 'carries' the wave, the way the medium moves, or the way the medium is disturbed as the wave passes an observer, is different. Sometimes, as with electromagnetic waves passing through the vacuum of deep space, there is no medium at all. However, despite these real differences there are many properties common to all these different types of wave. All waves, for example, are refracted, diffracted and reflected. All waves can interfere with each other, and many can be polarised as well.

These common properties can largely be explained in an uncomplicated way by the use of the Simple Harmonic Motion (SHM) mathematical model of a basic wave. This model usually deals with waves with the introduction of the so-called sinusoidal *wave equation*, which describes mathematically the varying shape of the common wave disturbances that are produced and the way in which these waves travel from location to location in terms of their frequency. This sine (or cosine) wave equation sets no preconditions on the medium carrying the wave, nor how the wave disturbs the medium in the first place, and it can be used for all kinds of waves. Repetitive particle motion requires practical limits (the extent of the back and forth motion about an equilibrium position) and a sine wave provides a good intuitive model to use. Importantly, sine wave model predictions agree extremely well with the common waves that we experience on a beach, as well as with waves detected by a wide range of non-human sensor systems in many different practical situations.

It is our intention in this short introduction to electromagnetic waves to review these common wave properties. As the title suggests, this book will look at several basic electromagnetic wave concepts that will become familiar to future marine engineers and physicists in their applications. More advanced discussion of electromagnetic waves and further reading of electrotechnology topics can be found elsewhere [I.1–I.4]. End of chapter self-assessment questions have been designed to cover a range of wave motion ideas to a level commensurate with the likely levels of the reader, with a difficulty level extending up to the first year of undergraduate physics and engineering studies. Finally, to get the most of this resource it is important to ask yourself what *your* expectations are from this book, your course, your tutor and yourself. Some of these questions are best addressed in the last chapter, which looks at teaching and learning skills issues.

References

[I.1] Reed's Marine Engineering and Technology Series: *Basic Electrotechnology for Engineers*, CR Lavers (Adlard Coles Nautical, 2013, ISBN 978–1-4081–7606–1).

[I.2] Reed's Marine Engineering and Technology Series: *Advanced Electrotechnology for Marine Engineers,* CR Lavers (Adlard Coles Nautical, 2014, ISBN 978–1-4081–7604–7).

[I.3] *Advanced Level Physics* (7th Revised Edition), M Nelkon and P Parker (Heinemann International Literature & Textbooks, 1995, ISBN 978–0-4359–2303–7).

[I.4] *Optics* (4th Edition), E Hecht and A Zajac (Addison Wesley, 2001, ISBN 978–0805385663; 3rd Revised edition, 1997, ISBN 978–0-2013–0425–1).

1

Basic Wave Concepts

'We ourselves feel that what we are doing is just a drop in the ocean. But the ocean would be less because of that missing drop.'   Mother Teresa

1.1 What is a wave?

A wave is usually regarded as the propagation of a disturbance, often periodic (i.e. repeating itself after a certain time), which travels through a medium or vacuum, transferring energy or information, and is accomplished without any permanent movement of the medium in the direction of the disturbance. The disturbance could be a single event, such as an underwater explosion; several waves, such as the 2004 South East Asian Boxing Day tsunami mentioned in the introduction; or a steady train of regularly emitted radar or sonar pulses.

In a wave, the energy of the vibration is usually moving away from the source in the form of a disturbance within the surrounding medium. However, in a standing wave on a string, energy can move in both directions equally. With ElectroMagnetic (EM) waves in a vacuum, the concept of a medium does not apply at all as there are no particles to physically transfer wave vibrations by 'bumping together'. Where there is a medium, wave energy is transferred by oscillations of the carrying medium; in the case of electromagnetic waves travelling through vacuum, the medium is considered to be the electromagnetic field of the wave itself.

We will look at the wave concept in several different ways to get the fullest understanding of what is taking place during the various wave motion processes encountered.

1.2 Wave parameters

In order to describe a wave, we need first to define some important fundamental parameters, which we can then use to quantify its description and appearance.

Consider first a cyclic motion, e.g. the motion of a small weight on the end of a string. As the particle rotates around the circle, the vertical and horizontal components of the radius change over time in a predictably repetitive manner (figure 1.1).

Figure 1.1: *Repetitive circular motion for radius vector R.*

The equation of a circle can easily be found by application of basic trigonometry to this particular geometry:

$x = Rsin\,\theta$ and $y = Rcos\,\theta$

Now:

$$x^2 + y^2 = R^2sin^2\theta + R^2cos^2\theta = R^2(sin^2\theta + cos^2\theta) = R^2$$

Since: $(sin^2\theta + cos^2\theta) = 1$

Figure 1.2 is a simple diagram showing the vertical displacement of the molecules of a medium from their original undisturbed position as a simple wave passes and an observer records the wave motion. The horizontal axis represents the distance from the source of the wave. Particles in air or water will vibrate about their *equilibrium position* with the *restoring force* provided by the lower density (lower pressure) region in figure 1.2 on the opposite side of the equilibrium position at any instant.

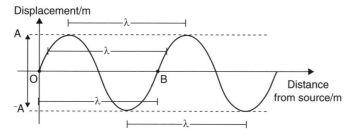

Figure 1.2: *Simple sine wave representation measured in terms of distance.*

The maximum displacement from the medium's undisturbed position is commonly called the wave *amplitude* (A). The amplitude can be both above the zero displacement (or equilibrium) line (+A), the wave *crest*, or below it (−A), the wave trough. In the example of figure 1.2, the wave amplitude can be measured in terms of centimetres. Amplitude is measured in Système International (SI) units of metres but is often significantly smaller than this for most electromagnetic wave applications – yet it can be much greater as well. Amplitude is often measured within radio and radar receiver electrical circuits as a detected voltage.

The distance between any two successive wave crests (or peaks), successive troughs, or between any two successive identical points on the waveform, known as the *wavelength* (λ – pronounced 'lambda'), is the distance travelled during one complete cycle (OB in figure 1.2) and can also be measured in SI units of metres, although wavelength can be much bigger or smaller than this. Successive points are those having the same displacement and the same slope (or gradient). In reality, the amplitude of a sound wave in water (or air) will decrease in amplitude over distance so the wave strength, and therefore energy carried, *dissipates* due to various losses and geometric spreading. Further wave dampening terms can be added to the Simple Harmonic Wave motion model but will not be explored here. For those interested in further study of this topic, there are many books that look at the damping of waves in some detail: [1.1–1.2].

An alternative way of viewing a wave is to show how the displacement of a particular point varies over time rather than with distance. The time taken for one complete cycle (vibration or complete oscillation) is called the *wave period*, or the time per cycle (T), and is measured in the SI units of seconds. However, most electromagnetic wave systems operate with wave periods much less than one second and so recording the period in seconds is not necessarily the most helpful measure of a wave's oscillatory behaviour.

It is generally more useful to quote the number of complete oscillations taking place in one second, which is found by taking the inverse of the repetitive periodic time, and this is called the *frequency* (f) of the wave or oscillation, i.e.

$f = 1/T$ **(eq 1.1)**

e.g., for a period of 0.02 seconds (or $T = 1/50$ s), the electrical frequency will be 50 Hz (the UK mains frequency). This relationship can be investigated a little further by first plotting the frequency as the period is decreased, for which a few sample values are given (Table 1.1).

Frequency (Hz)	Period T (seconds)
0.002	500
0.004	250
0.006	166.6667
0.008	125
0.01	100
0.012	83.33333
0.014	71.42857
0.016	62.5
0.018	55.55556
0.02	50

Table 1.1: *Frequency against period.*

Representing these values using a suitable software package such as Microsoft Excel will yield a graph similar to that in figure 1.3, which shows this frequency-period relationship.

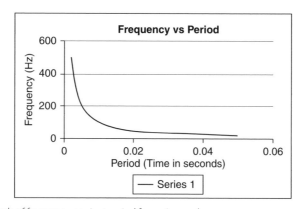

Figure 1.3: *Graph of frequency against period for various values.*

So how would the passage of a sine wave look over time? Actually, it would look remarkably similar to the representation already given in figure 1.2 as it is the same identical waveform we are considering, but simply measured in two different ways, and is given here in figure 1.4. However, we always plot period T against time and wavelength against distance and try to avoid mixing period with wavelength on the same graph.

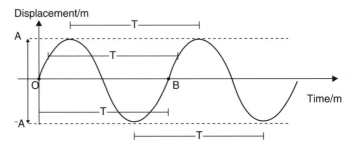

Figure 1.4: *Simple sine wave representation measured in terms of time.*

The unit of frequency is called the *hertz* (Hz) and is named after Heinrich Rudolf Hertz (1857–94), who first verified the existence of electromagnetic waves in 1887. Hertz was a German physicist at the University of Karlsruhe who expanded the electromagnetic theory of light first proposed by James Clerk Maxwell (1831–79) in 1865. Maxwell's most important achievement is regarded as formulating classical electromagnetic theory, which unified many previously unrelated observations, experiments and equations of electricity, magnetism and optics into a consistent, more comprehensive theory [1.3].

Maxwell's equations demonstrated that electricity, magnetism, heat and light are all in fact manifestations of the same phenomenon, namely the electromagnetic field. Subsequently, all other laws or equations of these subjects have now become simplified cases of Maxwell's more general equations. Heinrich Hertz was the first scientist to clearly demonstrate the existence of these electromagnetic waves by building a small radio device to produce and detect what we would now call Very High Frequency (VHF) or Ultra High Frequency (UHF) radio waves [1.4].

Most electromagnetic wave transmission involves very many wave vibrations per second. A frequency of 10 megahertz (10 MHz) is one where the oscillations occur at a rate of 10 million per second, and would be typical of analogue or Continuous Wave (CW) television broadcasts. A lower frequency of 5 kilohertz (5 kHz) is one where oscillations occur at a rate of 5000 per second and is typical of audio frequencies commonly found in human speech. Navigation radar commonly operates in the range 8–10 gigahertz (8–10 GHz) or 'India Band', some 8–10 thousand million vibrations per second.

1.3 Wave speed

Another important parameter is the **wave speed (V)**. Wave speed is the speed at which the wave travels through a medium. The standard unit for speed is metres per second (ms^{-1}), but km s^{-1}, km per hour, miles per hour or knots (nautical miles per hour) are commonly used alternatives. Wave speed depends upon: the wave type, the medium through which it is travelling and sometimes upon the wave frequency. If the direction as well as the magnitude of the speed is known, it is more usual to talk in terms of the wave velocity as this is a *vector* quantity, having both magnitude and direction. With reference to figure 1.5, consider R as a vector quantity. Vector R has a vertical component of $R \sin \theta$ and a horizontal component of $R \cos\theta$. Consider a vector aligned along the horizontal axis so the horizontal component of $R \cos \theta = R$ (since $\theta = 0$), while the vertical component $= R \sin \theta = 0$ (since $\theta = 0$ as well). Thus a vector quantity has no component of the vector at right angles (90 degrees) to it.

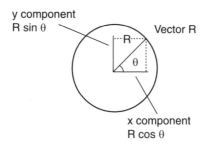

Figure 1.5: *Rotating vector calculation.*

1.4 Relation between speed, frequency and wavelength

The wave speed, V, is given by:

$$V = \frac{\text{Distance travelled}}{\text{Time taken to cover this distance}}$$

Consider a world-class sprinter who covers a distance of 100 m in a time of 10 seconds. The runner will have an average speed of 10 ms⁻¹, or an average velocity of 10 ms⁻¹ along the known direction of the running track. Note the speed at the start line is initially zero and changes rapidly, initially resulting in not only a rapidly changing velocity but a rapid change in acceleration (rate of change of velocity) as well. In fact, we can look at the actual 100 m statistics of Usain Bolt in the 2008 Olympics (figure 1.6).

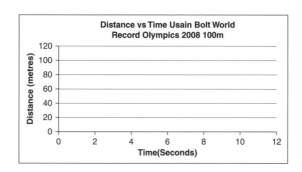

Figure 1.6: *2008 Olympic 100 m final distance vs time.*

Usain Bolt doesn't even start to move until 0.165 seconds after the gun is fired! On inspection of the graph, there appear to be two straight line regions joined together around the 2-second marker. An initial short section of variable and lower gradient is followed by a second section of higher constant gradient.

Now, from figures 1.2 and 1.4 we have already commented that the actual waves shown are identical; it is simply the way we measure or look at the wave that is different. Consequently, the time taken for a wave to travel one wavelength (e.g. from O to B on figure 1.4) is just the periodic time T, i.e. the time taken for one point on the wave to go through a single complete oscillation cycle.

So, $V = \dfrac{\lambda}{T}$

and since $f = \dfrac{1}{T}$, this may be rewritten as:

$V = f\lambda$ (**eq 1.2**)

This is an extremely important relationship, for if we know any two of speed (velocity, V), frequency and wavelength, we can easily calculate the third unknown quantity. This equation applies to all forms of wave motion.

Example 1.1: For a radar frequency of 3 GHz, what is the value of the wavelength (1 decimal place)?

Using $V = f\lambda$ and rearranging the equation for the wavelength

$\lambda = \dfrac{V}{f}$ and substituting for the values given, so: $\lambda = \dfrac{3 \times 10^8}{3 \times 10^9} = 0.1\ m$

In the unique case of electromagnetic waves, the speed of the wave is the same irrespective of the frequency of the wave and is extremely high; it is often referred to as 'c' after the Latin word *celeritas*, meaning swiftness. So for electromagnetic waves this equation becomes:

$c = f\lambda$ (**eq 1.3**)

Note: It is my view that Sir Isaac Newton's (1643–1727) choice of 'c' as the symbol to represent the swiftness of these waves may also reflect a pun on the word 'see', as the only electromagnetic waves known to man at that time were the visible ones!

It is valuable to remember some typical speed values at the start of this book:

sound speed in air at sea level	
(under standard atmospheric conditions)	about 330 ms⁻¹
sound speed in water	about 1500 ms⁻¹
light speed c	about 3×10^8 ms⁻¹

You may ask: 'What is Standard Atmospheric Pressure (STP)?' Well, it is an internationally accepted standard, i.e. STP is defined as a temperature of 273.15 kelvin (273.15 K ≡ 0 °C ≡ 32 °F) and an absolute pressure of 100,000 Pascals (Pa) equivalent to 1 bar pressure ≡ 14.5 pounds per square inch. Other useful data, standard engineering multiplier notation and commonly used Greek symbols are given in Appendix A.

The speed of sound in air is observed to fall with increasing height as the density (and thereby pressure) of air molecules falls, and hence the probability of molecules directly interacting with one another also decreases. The increased density of water molecules compared with air is the primary reason for the speed of sound in water being about five times greater than the speed of sound in air. It should also be appreciated that the speed of light provides the most rapid interrogation possible of the local above-water environment, making it suitable for radar sensing applications (where RADAR is an acronym for Radio Aid for Detection and Ranging). However, the poor transmission property of electromagnetic wave propagation in water (high absorption and strong scattering) generally renders the use of electromagnetic waves below water fairly useless except in the case of high-power blue/green underwater laser applications (which have low absorption and moderately low scattering).

Most radars operate using the *echo principle*, where electromagnetic wave pulses strike a target and a very small proportion of the emitted radar energy is reflected back towards the radar receiver (figure 1.7).

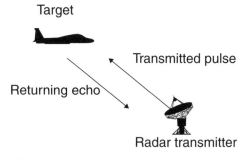

Figure 1.7: *Radar pulse delay ranging.*

If the relatively short elapsed time t, from transmission of the radar pulse to the detection of the received echo, is measured electronically, one can calculate the contact range as all electromagnetic waves travel at the same speed of light, $c = 3 \times 10^8$ ms^{-1}. Consequently, the distance they travel is given by:

Distance = ct and the target or contact range R is given by half the total distance travelled by the electromagnetic waves returning, so that:

Range R = ct/2   (**eq 1.4**)

Example 1.2: If the elapsed time of a long-range search radar pulse is 3 milliseconds, what is the range of the contact (3 significant figures)?
Range $= 3 \times 10^8 \times 3 \times 10^{-3}/2 = 4.5 \times 10^5$ m $= 450$ km

---

This relationship is used in pulsed radars to measure contact range and is known as pulse delay ranging, but applies to other forms of radiated EM energy such as LIDAR (LIght Detection And Ranging) as well as radiated sound energy underwater or SONAR (SOund Navigation And Ranging).

Lidar is now a very common airborne remote sensing technique used to produce swathes of millions of data points. From this, detailed digital terrain models can be mapped and analysed. Lidar surveying can safely collect data in inaccessible coastal areas, including soft cliff and saltmarsh areas, where other survey methods are unsuitable.

There can be significant deviation from the nominal values given for the different wave speeds in the above water and underwater environments, depending upon precise environmental conditions. For example, for propagation of sound waves in water or sonar waves, speed depends strongly upon the interplay of several related factors: salinity, temperature and pressure. It should not be surprising, therefore, that propagation speed in a warm, very salty environment, e.g. near the surface of the Dead Sea (figure 1.8), will be quite different to that in cold Arctic or Antarctic surface waters.

In oceanography, salinity refers to the water's saltiness. Salinity is not expressed as a percentage, but in parts per thousand (‰), or approximately the grams of salt per kilogram of aqueous solution. Optical refractive index, a property we will consider later, changes as the salinity changes, and also as the temperature changes. Salinity variation in the oceans of our planet are given elsewhere [1.5] and refractive index as a function of visible and infrared light is also given [1.6]. An example of refractive index change with temperature and salinity is given in Table 1.2.

Temperature	Salinity 30‰	Salinity 35‰	Salinity 40‰
0 °C	1.3398	1.3407	1.3419
10 °C	1.3393	1.3401	1.3412
20 °C	1.3384	1.3393	1.3402

Table 1.2: *Refractive index N of water with changes of temperature and salinity.*

Figure 1.8: *Dead Sea © CR Lavers.*

A more detailed description of the index of refraction of seawater and its dependence upon salinity, temperature and pressure can be found in the work of Austin and Halikas [1.7]. Other workers have also determined an empirical equation for predicting the absorbing component of refractive index (which, when expressed as a complex number, is referred to as the 'imaginary' index of water refraction [1.8]. Other workers have determined an empirical equation for the index of refraction of water as a function of temperature, salinity and wavelength at atmospheric pressure based on Austin and Halikas' work [1.9]. Further important water properties and parameters are introduced in Appendix B.

In a similar way, when light travels from air to enter a different medium, such as glass or water, both the speed and wavelength of light will be reduced, while the frequency remains unchanged. The pulse of light or *photon* cannot suddenly appear to gain energy as if from nowhere and retains its original energy level, which is dependent upon frequency. The refractive index N describes this effect with reference to the speed of light in vacuum,

$N = c/v$ (**eq 1.5**)

where c is the speed of light as discussed previously and v is the speed of light in vacuum in the new relevant medium.

Thus in glass, where the speed is (2/3)c typically, then $N = c/(2/3)c = 3/2 = 1.5$.

As light travels at approximately 300,000 kilometres per second in vacuum, which has a refractive index of 1.0 by definition ($N = c/c = 1$), it will slow down to about 225,000 kilometres per second in water (refractive index = 1.3) and drop further to 200,000 kilometres per second in glass (refractive index = 1.5). However, in diamond, with a relatively high refractive index of 2.4, or other high refractive index transparent materials, the speed of light can be reduced even more to about 125,000 kilometres per second, much less than its maximum speed in vacuum.

Example 1.3: A light wave travels in a transparent glass block at a speed of 0.6c. What is the glass refractive index (2 decimal places)?

Use $N = c/v$

Substituting: $N = c/(0.6c) = 1/0.6 = 1.67$

Note: It is possible for energetic electrons (or β^- 'beta' particles) to be produced from nuclear decay reactions underwater, along with visible and invisible radiation. These electrons can travel very close to the speed of light in vacuum and in fact greater than the speed of light in the water. This so-called Cerenkov radiation has been observed experimentally and may prove to be a future means of detecting nuclear materials and old nuclear submarines, especially if sensors are mounted on appropriate earth-viewing Low Earth Orbit (LEO) satellite-based sensor systems. Refraction will be covered in much greater detail in Chapter 4.

1.5 Phase and phase difference

Two other important wave parameters to consider are *phase* and *phase difference* (φ). Phase describes the point reached by an oscillation relative to another second reference point on the same wave in terms of angle. If the start of the actual wave oscillation is used as the reference point, then the phase of any point on that wave is measured by how far through one oscillation that point has reached, usually after a certain time interval has elapsed. The reader should be aware that phase cannot uniquely describe points that are more than one wavelength apart as each cycle of a wave contains an identical set of repeated displacement points. Consequently, 0° ≡ 360° ≡ 720° etc and the *angle clock* is reset after each complete cycle (figure 1.9). However, *phase difference* is used to describe effectively how one wave oscillates relative to another wave. This is very important when we have to consider the effect of combining two or more waves from sources together, and turns out to provide more useful insights than just considering time or distance, with which we are more familiar in our daily life.

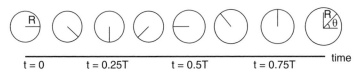

Figure 1.9: *Rotating clock.*

Each of the next three consecutive figures shows two waves of identical amplitude and frequency but with their peak values not occurring at the same time (figure 1.10) or the same place (figure 1.11). These curves can be represented mathematically, generally from the equation:

y = Asin ($\omega t - kx$) or more simply:

y = Asin ($\theta - \varphi$) which for a wave of unit amplitude simplifies further to:

y = sin ($\theta - \varphi$)

plotting a graph of y, the vertical displacement against θ where φ is a constant angle and both θ and φ are measured in degrees) is shown in figure 1.12. where φ is pronounced 'theta' and φ is pronounced 'phi'.

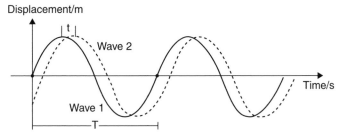

Figure 1.10: *Two waves with a time difference or time delay t between them, with Wave 1 leading Wave 2 in time by t seconds.*

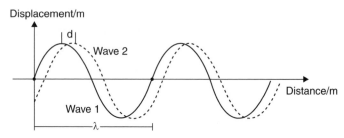

Figure 1.11: *Two waves with a path difference or path delay d between them, with Wave 1 leading Wave 2 in distance by d metres.*

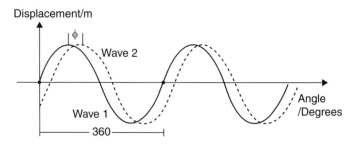

Figure 1.12: *Two waves with a phase difference or phase delay φ between them, with Wave 1 leading Wave 2 in angle by φ degrees.*

The dotted curve is shifted to the right relative to the solid curve by a time 't' in figure 1.10, by an equivalent path difference 'd' in figure 1.11 and by an equivalent phase angle 'φ' in figure 1.12. Phase difference can also be expressed as an angle in degrees but can be considered in *radians* where $360° \equiv 2\pi$ radians.

For these three representations to be equivalent, the fractional changes in each 'frame of reference' must be the same, i.e.:

$$\frac{t}{T} = \frac{d}{\lambda} = \frac{\phi}{360} \qquad \textbf{(eq 1.6)}$$

This useful relation between path difference and phase difference will be discussed further in the chapter detailing *interference* (Chapter 7).

Example 1.4: For a period of 3 seconds and a time delay of 0.5 seconds, what will be the equivalent phase shift in degrees between two waves arriving, having the same amplitude and frequency (2 significant figures)?
Using the equivalence relationship:

$$\frac{t}{T} = \frac{\phi}{360}$$

So $\phi = \frac{t}{T} \times 360 = 60°$

Note: In each of these three cases, the two waves are identical as they are the actual waves we are observing (or detecting). In each of the three figures we look at the waves in a different way, measuring a different property of the waves, of which there may be several (figures 1.10 to 1.12).

When there is no phase shift between two waves of identical amplitude and frequency, the waves are said to be *in-phase* – the waves exactly overlap and add together constructively, resulting in a wave of twice the original amplitude.

In modern phased array radar, the output of multiple small power sources is combined through the use of modern computer digital signal processing to produce a powerful narrow radar beam in a chosen direction, such as from the Royal Navy's new SAMPSON Radar fitted to the Type 45 Destroyer class.

However, when the peak of one wave Series 1 occurs at the same time (or in the same place) as the trough of a second wave Series 2 (but identical in frequency and amplitude), the resulting waves Series 3 are exactly half a waveform out of step. The waves are then said to be in *anti-phase* or 180 degrees out of phase (figure 1.13).

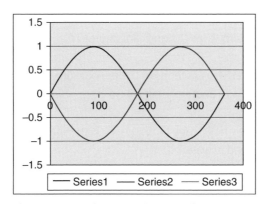

Figure 1.13: *Example of destructive interference combination of waves (Series 3 = Series 1 + Series 2).*

Such waves combine *destructively* and cancel each other out. Modern phased array radars produce almost complete cancellation in all directions except the chosen beam direction. By changing the phase conditions, it is possible to produce a constructive beam first in one direction and then in another. Active noise cancellation systems in aviation headphones work on the same principle. The sounds (frequencies and amplitudes) in a given space are monitored. Over time, many of these frequencies will exhibit a repetitive nature. The space within an entire airframe (fixed wing or helicopter) is too large to produce suitable cancellation. However, the small space within the earphones is suitable, producing cancelling noise dampening effects (typically -30dB) for the benefit of aircrew. Noise is thus deliberately produced and added in anti-phase to the actual system-generated noise, but now having components of equal and opposite magnitude that will cancel each other out.

Much more detailed material and discussions of mathematical principles relevant to this subject, including vectors, scalars, static and dynamic motion, work and power, are to be found in chapters 1–3 of Reeds Marine Engineering and Technology Series, Volume 2: *Applied Mechanics for Marine Engineers* (ISBN 978–1-4729–1056–1).

1.6 Self-assessment questions

After studying this chapter, you should be able to answer the following questions.

1.   Explain the terms: period, frequency, wavelength, amplitude and phase.

2.   Find the wavelength of the following electromagnetic waves:
        a. 9 GHz radar (3 decimal places)
        b. 85 MHz radio (3 decimal places)
        c. 10^{12} Hz microwaves (1 significant figure)

3.   Find the time delay t in terms of the periodic time T for a wave that is 45 degrees out of phase with another identical wave.

4.   Approximately how many times greater is the speed of light compared with the speed of sound in water?

5.   A lightning bolt is seen to strike the topmast of a yacht in a harbour. If the sound of the thunder is heard 8 seconds later, how far away is the yacht (1 decimal place)?

6.   Find the frequency for a sound wave in water travelling at a nominal speed of 1520 metres per second and having a wavelength of 3 cm (1 decimal place).

7.   In the United States of America, mains frequency is 60 Hz. What is the corresponding wave period (1 decimal place)?

8. $V = f\lambda$. Describe mathematically the change in speed with time as a function of both frequency and wavelength.

9.   If the group velocity is given by $V_g = \dfrac{\omega}{k}$ and the phase velocity is given by $V_p = \dfrac{d\omega}{dk}$, show how the phase velocity can be written in terms of the group velocity.

10.   If $V_g = V_0 e^{-kx-\omega t}$, find V_p in terms of V_g

REFERENCES

[1.1] *Introduction to Optics 3rd Edition*, F L Pedrotti, L M Pedrotti and L S Pedrotti (Pearson, 2006, ISBN 978–0-1319–7133–2).

[1.2] *The Physics of Vibrations and Waves 5th edition*, J Pain (Wiley Blackwell, 2005, ISBN 978–0-4700–1296–3).

[1.3] 'A Dynamical Theory of the Electromagnetic Field', J Clerk Maxwell, *Philosophical Transactions of the Royal Society of London* 155 (January 1865), pp. 459–512.

[1.4] *Electric waves: Being researches on the propagation of electric action with finite velocity through space*, Heinrich Hertz (1893; republished Dover Publications, New York, 1962).

[1.5] *Salinity Patterns in the Ocean* (Volume 1), L D Talley, 'The Earth system: physical and chemical dimensions of global environmental change' (pp 629–640), edited by Dr M C MacCracken and Dr J S Perry in *Encyclopedia of Global Environmental Change*, T Munn (editor) (Wiley, 2002, ISBN 978–0-4719–7796–4) http://www-pord.ucsd.edu/~ltalley/papers/2000s/wiley_talley_salinitypatterns.pdf

[1.6] 'Relative Reflectance and Complex Refractive Index in the Infrared for Saline Environmental Waters', M R Querry, W E Holland, R C Waring, L M Earls and M D Querry, *Journal of Geophysical Research*, Vol 82. No.9 (March 1977), pp. 1425–1433.

[1.7] 'The Index of Refraction of Seawater', RW Austin and G Halikas (Corporate author: Visibility Laboratory of the Scripps Institution of Oceanography, La Jolla, CA, January 1976) www.dtic.mil/cgi-bin/GetTRDoc?AD=ADA024800.

[1.8] 'Refractive index of water', www.philiplaven.com/p20.html.

[1.9] 'Empirical equation for the index of refraction of seawater', X Quan and E S Fry, *Applied Optics*, Vol. 34, Issue 18 (1995), pp. 3477–3480. http://dx.doi.org/10.1364/AO.34.003477.

2

Types of Waves

'This also is yours, the offering of their gift, even all the wave offerings of the sons of Israel'   Numbers 18:11

2.1 Different types of waves

Waves can have very different meanings; in the same way, it is possible to transmit different types of wave through all manner of materials: solids, liquids and even gases. For example, seismic waves can travel through the stratified layers of planet Earth, and mechanical waves will occur in solid structures such as bridges, buildings, wires, metal plates and ship bulkheads. Sound waves, meanwhile, can also travel through gases, fluids and solids. Considerable effort must be taken in the case of naval vessels to prevent sound energy being transmitted directly to the bulkhead or through the pressure hull of a submerged submarine, as sound detection can be a submarine's greatest enemy. These waves all have one factor in common: a material medium is necessary for wave transmission to take place, whereas only electromagnetic waves can travel through a vacuum.

2.2 Propagation of longitudinal sound waves

Sound waves may be produced by almost any form of energy, e.g. in a chemical explosion, through mechanical motion such as rubbing (friction), electrical spark, lightning, etc. Sound waves, however, cannot travel through the vacuum of space, for they are carried by the molecular collisions of the air particles, and in the vacuum of space there are no such molecules present. Sound waves can travel in other gases besides air, and they can travel in liquids or through solid bodies. Climbing to the top of Mount Everest or Mount Kilimanjaro, you would notice that the speed of sound in air falls considerably, as would the pressure. At sea level, standard atmospheric pressure is taken as about 1 bar or 1000 mbar. At the top of Everest, the 'weight' of atmosphere pressing down on a climber's head from above can drop below 500 mbar. In severe cooling weather conditions this can become a serious threat to the safety of climbers, who may then find themselves inadvertently above the limits of breathable oxygen as the atmosphere can contract down below the height of the climbers.

Consider the tuning fork in figure 2.1. Initially, in the undisturbed medium the air molecules are equally spaced as denoted by the spacing of the vertical lines at the top of the diagram (a). Once the tuning fork is struck and the fork compresses the air molecules, the local density is increased very slightly (several parts per million) and the wave begins to propagate into the air on the second line (b). A little while later, the arms of the tuning fork have begun to recoil in the opposite direction, back towards the original equilibrium position and a little beyond, while the compression is now clearly seen to have moved to the right (c). As the tuning fork continues to vibrate, the space between the arms is slightly increased and the air molecules are now slightly further apart on average than in the undisturbed medium, corresponding to a very slight decrease in the local density (*rarefaction*) of the air molecules, appearing as a slight increase in the spacing of the vertical lines (d). Over time, an alternating pattern of compressions followed by rarefactions begins to develop. This so-called pressure density wave is what causes changes in pressure across the eardrum when we hear a sound or a whale hears another whale underwater. The quietest sound we can just hear corresponds to the smallest variation (difference) in pressure across the membrane of the eardrum that we can sense.

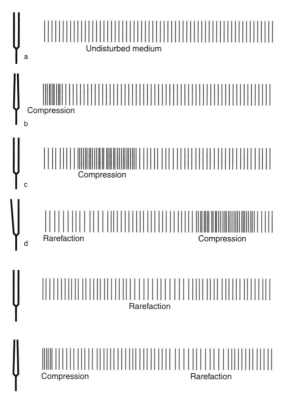

Figure 2.1: *Longitudinal wave generated with a tuning fork propagating as a series of compressions and rarefactions.*

Further discussion of acoustic sonar waves will be given in Chapter 8 but we will finish here by considering sound in pipes.

2.3 Standing waves and sound in pipes

These waves are similar to the natural periodic or standing waves at sea, or *seiches*, where waves bounce back and forth between the sides of a closed basin. Referring back to figure 1.2, points O and B are points of zero displacement, which appear 'fixed'. These so-called nodes are points along the zero displacement line about which a surface oscillates in the centre of the basin. An *antinode* is a point where there are maximum displacements of the surfaces as they oscillate, i.e. a wave peak (wave amplitude of maximum displacement). A *basin*, like an organ pipe, can be closed or open. Resonance amplifies the displacement at the antinodes and occurs when the period of the basin is similar to the period of the force producing the standing waves.

It is quite easy to create a stationary resonant harmonic sound wave by blowing over a bottle or by playing an organ. Let us consider closed and open pipe ends. Obviously air cannot penetrate a closed pipe end. Air molecules at the end of a closed organ pipe are thus stationary or fixed; they are unable to move or be displaced into or out of the closed end. The closed end of a pipe is thus a node or displacement node. However, to not displace air the closed pipe end must exert a force on the molecules by means of pressure, so the closed end is a pressure antinode.

At an open pipe end, the argument is reversed. The pipe is open to air so there must be a pressure node at the open end. Pressure and displacement are 90 degrees out of phase, so the open end is a displacement antinode.

In practice, the air pressure in a standing wave doesn't instantly equalise with the background pressure at the open end – it extends out of the pipe a little way. The displacement antinode is thus just a little outside of the pipe end and not at the pipe end, so open organ pipe wave shapes are a bit 'longer' than they appear from the point of view of computing their resonant harmonics.

2.3.1 Pipe open at both ends

There are displacement antinodes at both ends as air molecules are free to move. This is, of course, just like a string unrestricted to move at both ends. A free string end is obviously free to whip up and down. Let us take a possible solution of the form:

$f(x,t) = A_0 \cos k_n x \cos(\omega_n t)$ **(eq 2.1)** and $\cos k_n L = \pm 1$

For $k_n L = n\pi$, remember there are *pressure* nodes at both ends, which makes them like a string fixed at both ends again (see figure 2.2).

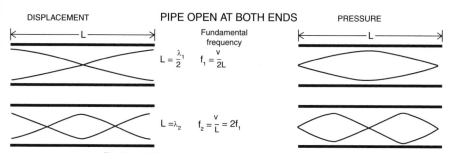

DISPLACEMENT   PIPE OPEN AT BOTH ENDS   PRESSURE

Fundamental frequency

$L = \dfrac{\lambda_1}{2}$ $f_1 = \dfrac{v}{2L}$

$L = \lambda_2$ $f_2 = \dfrac{v}{L} = 2f_1$

First harmonic frequency = twice the fundamental frequency

Figure 2.2: *Displacement and pressure variations for a pipe open at both ends.*

2.3.2 Pipe closed at one end

There is a displacement node at the closed end, and an antinode at the open end. This is just like a string fixed at one end and free at the other. Let's arbitrarily make x = 0 the closed end. Then:

$f(x,t) = A_0 \sin k_n x \cos(\omega_n t)$ **(eq 2.2)** and having a node at x = 0 for all possible k. To get an antinode at the opposite end, it is necessary that $\sin k_n L = \pm 1$.

Then to get odd half-integral multiple values of π for $k_n L$ for every value of n, we require that:

$k_n L = \dfrac{2n-1}{2}\pi$ for n = 1,2,3, etc **(eq 2.3)** as the shortest length of the pipe to get an antinode and node present will have a length of a quarter wavelength.

Or rather, rearranging for k:

$k_n = \dfrac{2n-1}{2L}\pi$ and remembering that $k = \dfrac{2\pi}{\lambda}$ Thus $\lambda = \dfrac{2\pi}{k}$ and so rephrasing **(eq 2.3)**

$\lambda_n = \dfrac{2\pi}{(2n-1)\pi}\dfrac{2L}{} = \dfrac{2\pi}{\pi}\dfrac{2L}{(2n-1)} = \dfrac{4L}{(2n-1)}$ **(eq 2.4)**

And given that $\lambda = \dfrac{v}{f}$

Thus:

$$f_n = \dfrac{(2n - 1)v}{4L}$$

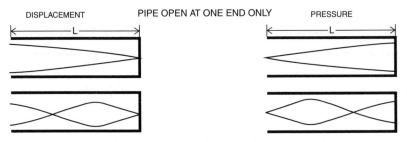

Figure 2.3: *Displacement and pressure variations for a pipe closed at one end.*

2.4 Propagation of transverse mechanical waves

Unlike longitudinal waves, this particular type of wave is associated with materials that have strong intermolecular bonds, such as plastics, wood, various types of metals, etc. Initially, particles are undisturbed (a), but then one is disturbed by a vertical motion (b) (figure 2.2). This first particle vibrates about the equilibrium position, as time progresses, out to a maximum extension (c). But this first particle is physically connected or joined to the second by molecular forces holding the material together and so a restoring force starts to act upon the first molecule, bringing it back from its position of maximum extension. Because of the molecular forces holding the material together, the second particle is forced to repeat the motion of the first but slightly later in time (d), and so on. Similar effects occur for all other particles in the chain. The disturbance thus travels through the medium until the disturbance has passed completely by the first particle, which once more returns to its undisturbed equilibrium position (i)- through a variety of loss or dampening mechanisms.

The particles thus move in a direction at right angles (orthogonal, or at 90 degrees) to the wave propagation direction, and is said to be a *transverse wave*. This type of wave motion can be demonstrated by giving a taunt string a flick at right angles to the direction of the string.

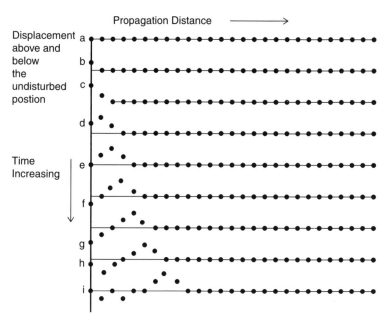

Figure 2.4: *Transverse wave generated along a string, showing individual molecular motion developing over time.*

As we have already seen, sound waves are longitudinal, while waves on a string are transverse. Many other forms of wave motion can be built up from these two fundamental modes. For example, the waves on the surface of the sea, which have a complicated form, may simply be considered as combinations of both longitudinal and transverse wave motion out of step, since if a particle moves forwards, then downwards, backwards and upwards, it will actually describe a rotational motion. Longitudinal and transverse waves exist throughout a medium, while the wave motion that causes waves on the surface of the sea is restricted to the surface layer, where the layer change occurs (or the break in symmetry of the system). As we go below or above the surface, the motion changes so there are no longer any rotational waves. Surface and sub-surface waves will be examined a little later in this chapter (in section 2.6).

2.5 Simple harmonic motion

Let us first consider mathematically the case of *Simple Harmonic Motion* (SHM) before moving on to consider more complex surface water waves. SHM occurs when the force on an oscillating object is proportional and in the opposite direction to the object's displacement. Common examples include masses on springs and pendulums, which 'bounce' back and forth repeatedly. Mathematically, this can be written:

$F = -ky(t)$ **(eq 2.5)**

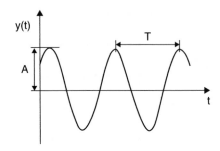

Figure 2.5: *Graph of displacement against time in simple harmonic motion.*

where F is force, $y(t)$ is the displacement measured with respect to time, and k is a positive constant. This is exactly the same as Hooke's law, which states that the force F on an object at the end of a spring equals: $-ky(t)$, where k is the spring constant. Since $F = ma$, and acceleration is the second derivative of displacement with respect to time t:

$$m\frac{d^2y}{dt^2} = -ky(t)$$ **(eq 2.6)**

Thus: $\frac{d^2y}{dt^2} = -ky(t) / m$ **(eq 2.7)**

The general solution of this type of second order differential equation is:

$y(t) = A\cos(\omega t) + B\sin(\omega t)$ **(eq 2.8)**

where the angular velocity of the object is $\omega = 2n f d$ and f is the frequency of the wave and n is an integer value.

The values of A and B are determined by the initial conditions. The general solution can also be written more compactly as:

$y(t) = A\cos(\omega t - \phi)$ **(eq 2.9)**, where ϕ represents any phase shift or delay between the vertical motion and the time.

2.5.1 Velocity and acceleration

Given that the displacement of a simple harmonic oscillator is:

$$y(t) = A\cos(\omega t - \phi)d$$

then the velocity, the rate of change of displacement, will be given by:

$$v(t) = \frac{dy}{dt} = A\omega\sin(\omega t - \phi) \qquad \textbf{(eq 2.10)}$$

and the acceleration, the rate of change of velocity, will be:

$$a(t) = \frac{dv}{dt} = A\omega^2\cos(\omega t - \phi) \qquad \textbf{(eq 2.11)}$$

Now, from equation (2.5), given that $y(t) = A\cos(\omega t - \phi)$, substituting for $A\cos(\omega t - \phi)$ into equation (2.7) gives:

$$a(t) = -\omega^2 y(t) \qquad \textbf{(eq 2.12)}$$

2.5.2 Angular velocity

Angular velocity in circular motion is the rate of change of angle. It is measured in radians per second. Now, since 2π radians is equivalent to one complete rotation in time period T:

$$\omega = \frac{2\pi}{T} = 2\pi f \qquad \textbf{(eq 2.13)}$$

Let's return to our equation for displacement in simple harmonic motion:

$y(t) = A\cos(\omega t - \phi)$ or $y(t) = A\cos(\omega t - kx)$, where x is the direction of the wave propagation and $k = \frac{2\pi}{\lambda}$.

The equation includes angular velocity in its simple harmonic motion and is similar to circular motion. A mass on a spring undergoes SHM. If you look at an object going round in a circle side-on, it will look identical to SHM.

Now: for motion in a circle $\omega = \dfrac{2\pi}{T}$ and

$a(t) = \dfrac{d^2y}{dt^2} = -A\,\omega^2\,y(t)$, while for a spring:

$\dfrac{d^2y}{dt^2} = -ky(t)\,/\,m$

Equating both versions of the acceleration.

Hence $-A\,\omega^2\,y(t) = -ky(t)\,/\,m$

So $A\,\omega^2 = k\,/\,m$

and $\omega^2 = k\,/\,Am$, and thus

$\omega = \sqrt{\dfrac{k}{Am}}$, which for a unit amplitude of A = 1 becomes $\omega = \sqrt{\dfrac{k}{m}}$ **(eq 2.14)**

Example 2.1: Consider a 1 N weight that extends a spring by 3 cm. A second 1 N weight is added, and the spring extends a further 3 cm. What is the spring constant (2 decimal places)?

$\Delta F = k\Delta x$

$1 = 0.03k$

So:

$k = \dfrac{1}{0.03} = 33.33$

Example 2.2: A pendulum oscillates with a frequency of 10 Hz. What is the length of the pendulum (2 significant figures)? For a pendulum $\omega = \sqrt{\dfrac{g}{l}}$

Now: $\omega = 2\pi f$ and for a pendulum $\omega = \sqrt{\dfrac{g}{l}}$ so $\omega = \sqrt{\dfrac{g}{l}} = 2\pi \times 10$

and thus $l = \dfrac{g}{\omega^2} = \dfrac{g}{(2\pi \times 10)^2} = l = 0.0025\ \text{m}$

2.6 Surface waves in deep and shallow water

There are two main categories of waves: *progressive waves* and *standing waves*. Progressive waves include: surface waves, internal waves and tsunamis. We have already discussed standing waves, like those found on a string or slinky spring. Progressive waves are those that move forwards across a surface. As a wave passes a waveform, energy moves forwards but individual water molecules orbit *radially*, increasing in radius with increasing wave size and decreasing with depth below the water surface.

2.6.1 Waves in deep water

Surface waves are produced when wind blows over a liquid surface. While the driving force of the wind continues, the wavelength and height of the waves created will increase with the overall distance travelled, up to a limit dependent on wind velocity. When the wind stops, waves continue with unchanged wavelength but gradually decrease in height, due to attenuation (or energy loss). A real sea surface is complex, and will have a wide spectrum of wavelengths present. Slight differences in the direction of component waves cause otherwise uniform crest lines to 'break' seemingly randomly after travelling large marine distances. The greater the distance over water that waves travel (the fetch), the higher the resulting waves. The highest waves generally occur in the Southern Ocean. Here, waves occur in excess of 6 metres in height, correlating with the world's strongest winds (which can be greater than 15 ms^{-1}). The lowest waves, by comparison, are found in the tropical and subtropical oceans, where wind speed is also the lowest.

To understand the wave motion processes involved in surface waves, we must first consider a *wave train* having constant amplitude and wavelength. It is useful to start with the assumption that the surface form of a self-maintaining wave is *cycloidal* (as shown in figure 2.6). In this regard, if you can get hold of this now out-of-print book it is very helpful to explain cycloidal motion and other water wave related issues that we do not go into in this basic book [2.1]. Each surface particle rotates in a circle of, let us say, radius A (thus the amplitude A is half the wave height), crest to trough. The phase of the rotation changes with distance along the wave train. Thus, if we take the positive direction x to be the direction of the wave motion and we define any particle by its x coordinate in the absence of wave motion, then in the presence of a wave an individual water molecule will find itself at the head of a radius vector of length A, making an angle θ with the forward horizontal direction, where:

$$\theta = -\omega t + 2\pi \frac{x}{\lambda} \qquad \textbf{(eq 2.15)}$$

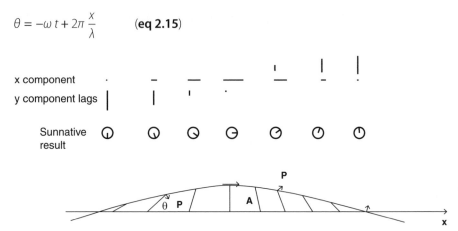

Figure 2.6: *Movement of surface particles in a progressive wave, showing the difference between x and y components and their summative interaction in the cycloidal motion.*

This cycloidal motion can be treated as the sum of two simple harmonic motions, one vertical with amplitude A, and one horizontal, also of amplitude A but differing in phase from the first vertical waveform by $\frac{\pi}{2}$, remembering that any vector at an angle to the horizontal will have both a vertical and a horizontal component. The vertical wave component lags the horizontal movement, which starts progressively ahead in the +ve x direction, i.e.:

$$x(t) = A \sin \theta = A \sin(kx - \omega t)$$

$$y(t) = A \sin\left(\theta + \frac{\pi}{2}\right) = A \sin\left(kx - \omega t + \frac{\pi}{2}\right)$$

2.6.2 Wavelength and phase velocity

The fundamental idea is that the water surface will set itself perpendicular (at right angles) to the resultant force. The resultant force on any surface particle is the combination of the force of gravity mg acting towards the centre of the Earth and the centripetal force $m\omega^2 A$ due to the rotary motion, giving rise to the general description of such a surface wave as a *gravity wave*.

It is very convenient to describe a specific point of interest on the wave surface, e.g. point P in figure 2.6, by the phase angle θ that corresponds to it. We can then determine the deviation from the vertical of the resultant force R by reference to figure 2.7, from which it is clear that using the two right-angled triangles appropriately:

Since $\tan a = \dfrac{opposite\ BD}{adjacent\ CE}$ and $BD = m\,\omega^2 A\cos\theta$ (as $\cos\theta = \dfrac{BD}{m\,\omega^2 A}$) and

$CE = mg - m\,\omega^2 A\sin\theta$

so

$$\tan a = \frac{m\,\omega^2 A\cos\theta}{mg - m\,\omega^2 A\sin\theta} = \frac{\omega^2 A\cos\theta}{g - \omega^2\sin\theta} = \frac{A\cos\theta}{\frac{g}{\omega^2} - A\sin\theta} \qquad \textbf{(eq 2.16)}$$

We next consider the wave slope at the same point P on the surface.

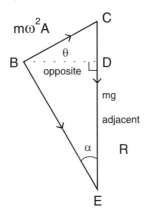

Figure 2.7: *Diagram of forces.*

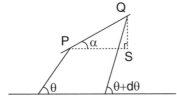

Figure 2.8: *Surface slope steepening between P and Q, a point infinitesimally further up the slope.*

Referring to figure 2.8, which represents the water wave profile at a point P, it is also observed that the vertical distance from P to Q is:

$y = A\sin\theta$

So: $y + dy = A\sin(\theta + d\theta)$

Hence:

$dy = A\sin(\theta + d\theta) - A\sin\theta$

And so:

$$\frac{dy}{d\theta} = \frac{A\sin(\theta + d\theta) - A\sin\theta}{d\theta}$$

$$\frac{dy}{d\theta} = A(d\sin\theta)$$

$$\frac{dy}{d\theta} = A\cos\theta$$

so: $dy = A\cos\theta \, d\theta$

The horizontal distance from P to Q is $dx + dx'$, where dx' is found by considering:

$x = A\cos\theta$ and

$x' = A\cos(\theta + d\theta)$

Hence: $dx' = x' - x = A\cos(\theta + d\theta) - A\cos\theta$

$$\frac{dx'}{d\theta} = \frac{A\cos(\theta + d\theta) - A\cos\theta}{d\theta}$$

$dx' = Ad(\cos\theta) = -A\sin\theta$

This is an extremely sensible answer, as $d\theta$ increases the point Q to tilt back further towards P.

Hence the horizontal distance from P to Q is:

$dx - A\sin\theta$

But $\theta = -\omega\, t + 2\pi\dfrac{x}{\lambda}$

So at any fixed time t, by differentiation we have:

$d\theta = 2\pi\dfrac{dx}{\lambda}$ or rearranging for dx we find that $dx = \lambda d\theta/2\pi$

Thus $tan\, a = \dfrac{SQ}{PS} = \dfrac{A\cos\theta\, d\theta}{dx - A\sin\theta d\theta} = \dfrac{A\cos\theta}{\frac{\lambda}{2\pi} - A\sin\theta}$ (eq 2.17)

Note: The topic of differentiation, or the method of instantaneously finding the slope or gradient for any general waveform, is discussed further in Appendix C.

If these two results are equated for all possible values of the phase angle θ, we must have the following equality:

$$\dfrac{A\cos\theta}{\frac{\lambda}{2\pi} - A\sin\theta} = \dfrac{A\cos\theta}{\frac{g}{\omega^2} - A\sin\theta}$$ (eq 2.18)

For these to be equal for all values of θ, $\dfrac{\lambda}{2\pi} = \dfrac{g}{\omega^2}$ (eq 2.19)

Now, the *phase velocity* of such cycloidal wave motion is given by $c = f\lambda = \dfrac{\omega\lambda}{2\pi}$, where c is the wave speed along the surface of the water and *not* the speed of electromagnetic waves in vacuum.

Such that: $\omega^2 = \dfrac{4\pi^2 c^2}{\lambda^2}$ (eq 2.20)

Substituting this value into equation (2.18), we obtain $\dfrac{\lambda}{2\pi} = \dfrac{g\,\lambda^2}{4\pi^2 c^2}$

So that $c^2 = \dfrac{g\lambda}{2\pi}$ or $c = \sqrt{\left(\dfrac{g\lambda}{2\pi}\right)}$

Under these conditions for phase velocity, a cycloidal surface wave can occur. For example, consider waves with a wavelength of 10 m; the phase velocity will be given by:

$$c = \sqrt{\dfrac{g\lambda}{2\pi}} = \sqrt{\dfrac{9.81 \times 10}{2\pi}} = 3.95 \text{ ms}^{-1}(2 \text{ decimal places}).$$

While there are other factors to consider concerning sea wave amplitude at any wavelength, there is an absolute limit given by the relationship: $A\omega^2 = \dfrac{2\pi g}{\lambda}$. If a wave exceeds this amplitude, crests become separated from the main body of the waves, so an individual wave cannot continue as a progressive wave any longer. This limit is found by substituting for $\omega^2 = \dfrac{2\pi}{\lambda}g$ (from equation 2.19 into 2.20), so that:

$A2\pi g / \lambda = g$ or $A_x = \lambda / 2\pi$

If the relevant cycloidal surface is drawn for this limiting case, it is noted the wave crests become peaks: i.e. there is a discontinuity in the wave slope at each crest (figure 2.9). Progressive waves with this amplitude and steepness are not found at sea. It is useful to consider such limiting amplitudes in the study of a disturbance that continues beneath the sea surface, which is associated with wind-driven surface waves. Additionally, when a wave crest travels faster than the wave beneath it, the waves start to break or collapse. Waves will break depending on the depth of the water body.

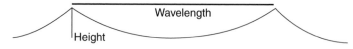

Figure 2.9: *Wave steepness.*

Wave steepness is the ratio of wave height H divided by wavelength (H/λ). In shallow water, H increases and λ decreases and thus wave steepness increases. Deep-water waves have heights greater than $\dfrac{\lambda}{4}$ and phase velocity = g/ω = gT/(2π)

$\lambda = cT = \dfrac{gT^2}{2\pi}$ or from $c = \dfrac{\lambda}{T}$ for a given period T.

Group velocity Vg $= g/(2\omega) = \dfrac{gT^2}{4\pi}$, which is of course half the phase velocity. For a given period, waves of longer wavelength travel faster.

2.6.3 Sub-surface wave motion

The previous discussion related the slope of the wave surface to the downward deflection of the resultant force. Now, we can apply this idea very generally to any equal pressure isobaric surface and not just to the more obvious air/water surface (which is, of course, just one specific isobaric surface among many). Hence we can examine other possible cycloidal waves on any sub-surface isobaric equal pressure region, and these equations are applicable to all sub-surface isobaric surfaces.

Let us consider how the amplitude of these cycloidal waves varies with depth. It is clear that the amplitude at a sub-surface isobar will be less than that at the surface, i.e. the surface water molecules have the greatest freedom to oscillate or 'undulate' up and down than those below the water surface. In addition, the driving force of the oscillation is to be found at this water/air boundary and not in the depths of the sea. Below the surface, the resultant force on any particle at any arbitrary wave crest is $m(g - \omega^2 A)$ in the downward direction.

From pressure considerations, there will be a change in pressure $dp = (g - \omega^2 A)\rho dz'$ over a height dz', so the rate of pressure increase with depth under a sub-surface crest is *less* than in undisturbed water.

Under a sub-surface trough $dp = (g + \omega^2 A)\rho dz'$, so the rate of pressure increases with depth will be *greater* than in undisturbed water.

The rate of pressure increase with depth for sub-surface waves is also greater than that for undisturbed water, as clearly a sub-surface wave will disturb the equilibrium conditions that would otherwise exist. So to reach the pressure corresponding to 1 m depth in undisturbed water, we must go down from a crest more than 1 m and to a trough less than 1 m. So the wave amplitude at the

1 m isobar is less than at the surface. If we consider more fully the situation under a crest, we have:

$$dz' = \frac{1}{\rho(g - \omega^2 A)}\, dp,$$ where $z\, \gamma$ is the depth from the crest to the p pressure isobar.

Given that pressure = force/area and considering a specific thickness dz:

In undisturbed water we have $dz = \dfrac{1}{\rho g}\, dp$

The difference between these values is the decrease in amplitude corresponding to the pressure change dp:

$$dA = -(dz' - dz) = -\frac{1}{\rho g}\, dp\left(\frac{1}{\left(1 - \frac{\omega^2}{g} A\right)} - 1\right)$$

Thus $\left(\left[\dfrac{g}{\omega^2 A}\right] - 1\right) dA = -\dfrac{1}{\rho g}\, dp$

So $\left(\left[\dfrac{g}{\omega^2}\right]\log\left(\dfrac{A}{A_0}\right)\right) - (A - A_0) = -\dfrac{p}{\rho g} + \dfrac{p_0}{\rho g}$ **(eq 2.21)**

Leading to $\dfrac{g}{\omega^2}\log\left(\dfrac{A}{A_0}\right) = -\dfrac{p}{\rho g} + \dfrac{p_0}{\rho g} + (A - A_0)$

and since $A = A_0$ for $p = 0$ at the sea surface and since $\dfrac{g}{\omega^2} = \dfrac{\lambda}{2\pi}$, we can rewrite equation 2.21 as:

$$\log\left(\frac{A}{A_0}\right) = \frac{\omega^2}{g}\left[0 - \frac{p}{g\rho} + (A - A_0)\right] = \frac{2\pi}{\lambda}\left(-\frac{p}{g\rho} + A - A_0\right)$$

Now: $\dfrac{p}{g\rho} = -z$

so $\log\left(\dfrac{A}{A_0}\right) = \dfrac{2\pi}{\lambda}(z + A - A_0)$ **(eq 2.22)**

We have so far been measuring from a surface crest to an isobar crest. If we instead measure from the centre of rotation of surface particles to the centre of rotation of particles in the p isobaric pressure surface, this becomes more simply:

$$log\frac{A}{A_0} = -\frac{2\pi z}{\lambda}$$

If we also assume that the surface wave is the maximum wave, we can put $A_0 = \frac{\lambda}{2\pi}$ and can then derive a general expression for the penetration of the wave disturbance into the sea, which is found by taking inverse logarithms so that:

$$\frac{A}{A_0} = \exp\left(-\frac{2\pi z}{\lambda}\right) \text{ or rearranging so } A = A_0\exp\left(-\frac{2\pi z}{\lambda}\right) \qquad \textbf{(eq 2.23)}$$

A similar argument can be made for the depth of isobars beneath a trough.

Figure 2.10 shows how wave amplitude decreases with depth with A_0 set to unit magnitude. This is very general, so we can take a surface level corresponding to any selected wave amplitude and then measure down in wavelength to find the depth at which the amplitude has decreased to a particular value (e.g. the e^{-1} value).

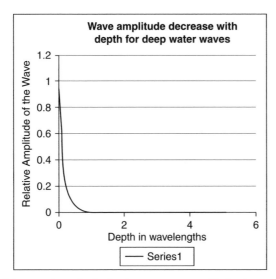

Figure 2.10: *Plot of decreased amplitude of a deep-water wave with depth.*

The layers below the sea surface show an appropriate decrease in sub-surface conditions by comparison with the sea surface, as depth increases until the motion finally diminishes until it is no longer detectable. It is clear that, regardless of the actual wavelength, the wave amplitude decreases substantially within the first wavelength depth. This is a good point to introduce the idea of the wave base, the depth to which surface waves can move water. If the water is deeper than the wave base, orbits are circular and there is no interaction between the bottom and the wave. However, if the water is shallower than the wave base, orbits are elliptical and are increasingly flattened towards the bottom, figure 2.11.

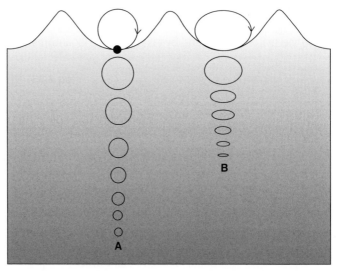

Figure 2.11: *A particle motion in an ocean wave. A = Deep water, B = Shallow water. The elliptical movement of a surface particle flattens with increasing depth. (en.wikipedia.org/wiki/File:Wave_motion-i18n-mod.svg Attributions)*

There are three types of waves defined by water depth: deep-water waves, where depth is much greater than the wavelength λ; intermediate depth, where $\frac{\lambda}{20} <$ depth $< \lambda$; and shallow water waves, where depth is less than $\frac{\lambda}{20}$.

Example 2.3: A storm at sea generates deep-water waves with a period of 30 s. If the storm is 500 km offshore, how long will it take the waves to get to shore?

Although crests travel at the phase speed, energy from the waves generated by the storm can only travel at the group speed.

Since for deep-water waves $c = \sqrt{\dfrac{g\lambda}{2\pi}} = \sqrt{\dfrac{g}{k}}$

As $c_g = \dfrac{d\omega}{dk} = \dfrac{g}{2\omega} = \dfrac{1}{2}\sqrt{\dfrac{g}{k}} = \dfrac{c}{2}$

$$C_g = \frac{9.81}{22\pi / 30} = 23.42 \text{ metres per second.}$$

The time to reach the shore will be given by the group velocity divided by the distance:

$$Time = \frac{500 \ km}{23.42 \ metres \ per \ second} = 5.93 \ hours.$$

In deep water, it is even possible to estimate the speed of a ship as there is a simple relation between the ship velocity v, the wave velocity c and the angle θ that the wavefront makes with the ship's path (figure 2.2):

$$v\sin\theta = c \qquad \textbf{(eq 2.24)}$$

Deep-water surface waves of wavelength λ have phase velocity given by: $c = \sqrt{\dfrac{g\lambda}{2\pi}}$,

where g is the acceleration due to gravity. Substituting this value into **eq 2.24**, we obtain

$$v = \frac{\sqrt{\frac{g\lambda}{2\pi}}}{\sin\theta} \textbf{(eq 2.25)} \text{ [2.2].}$$

position time t_2

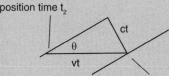

Figure 2.12: *Boat wave front calculation.*

Figure 2.12. Section of a wave front in the wake of a large ship, shown at two times: t_1 and t_2. The ship moves from right to left with velocity v. The phase velocity of the wave is c, and the angle between the wave front and the ship's path is θ.

Example 2.4: Find the estimated speed in metres per second of a vessel image taken by satellite with wavelength of 20 m and an angle between the vessel's path and the wave front of 40 degrees (2 decimal places).

Using $v = \dfrac{\sqrt{\frac{g\lambda}{2\pi}}}{\sin\theta}$ and substituting $v = \dfrac{\sqrt{\frac{9.81 \times 20}{2\pi}}}{\sin 40} = 8.69$ metres per second

2.7 Waves in shallow water

Let us now consider an extreme situation, where the water depth z is instead much less than the wavelength. For simplicity, we will also assume that the wave amplitude A is small and much less than the wavelength (a generally reasonable assumption). In such circumstances, the phase velocity will follow a very different law than that for deep-water waves, and being dependent only on depth will thus be independent of the wavelength under these conditions.

In the case of waves in shallow water, Bernoulli's hydrodynamic flow theorem is needed. For streamlined flow, we consider the conditions applying to flow tubes, i.e. tubes supposed in our argument to be parallel to the streamlines. We assume that there are no frictional losses at the edges, so the work done on the water in the tube will be exactly equal to the energy gained by the water. This energy consists of both kinetic and potential energies.

Let a small volume n flow through the tube in the direction shown in figure 2.13.

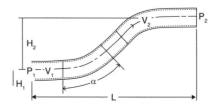

Figure 2.13: *Small volume flow.*

Then we have:

Work done on the system $= p_1 \Delta$

Work done *by* the system $= p_2 \Delta$

So net work done on a system $= p_1 \Delta - p_2 \Delta$

The corresponding mass of a volume Δ is $\rho \Delta$ where ρ is the density

Gain in kinetic energy $= \dfrac{1}{2} \rho \Delta v_2{}^2 - \dfrac{1}{2} \rho \Delta v_1{}^2$ where v_1 and v_2 are the initial and final velocities respectively.

Gain in potential energy $= \rho \Delta g H_2 - \rho \Delta g H_1$

(*consider potential energy of a mass m raised thorugh height h mgh*)

Thus, since the work done on the system = the gain in the energy of the system

Then the work done on the system = the gain in the energy of the system

So that: $p_1 \Delta - p_2 \Delta = \frac{1}{2}\rho \Delta v_2{}^2 - \frac{1}{2}\rho \Delta v_1{}^2 + \rho \Delta g H_2 - \rho \Delta g H_1$

and by sorting and cancelling the terms on both sides of the equation we obtain:

$$\frac{p_1}{\rho} + \frac{v_1{}^2}{2} + gH_1 = \frac{p_2}{\rho} + \frac{v_2{}^2}{2} + gH_2$$

Or more simply:

$$\frac{p}{\rho} + \frac{v^2}{2} + gH = constant \qquad \textbf{(eq 2.25)}$$

Let us now consider the application of this general theorem to our water surface wave motion. Consider water flowing left to right under the wave profile at mean velocity c. It will pass through a shallower cross section under a trough than it does under a crest and so its velocity will be higher under the trough. If the velocity under the trough is $c + v$, it will therefore be $c - v$ under the crest.

From figure 2.14 it can be seen that $\dfrac{c+v}{c-v} = \dfrac{z+h}{z-h}$

Hence: $(c + v)(z - h) = (z + h)(c - v)$

And $cz - ch + vz - hv = cz - vz + ch - hv$

Which after simplification becomes: $2vz = 2ch$

Or $\dfrac{c}{v} = \dfrac{z}{h}$

So now applying Bernoulli's theorem to a tube of flow in the surface layer from crest to consecutive trough:

$$\frac{p}{\rho} + \frac{(c+v)^2}{2} + g(z-h) = \frac{p}{\rho} + \frac{(c-v)^2}{2} + g(z+h)$$

Thus:

$$\frac{p}{\rho} + \frac{(c^2 + v^2 + 2cv)}{2} + gz - gh = \frac{p}{\rho} + \frac{(c^2 + v^2 - 2cv)}{2} + gz + gh$$

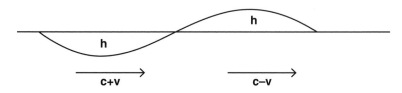

Figure 2.14: *Shallow-water wave motion with local decrease and increase in pressure and consequent local increase and decrease in flow speed under wave trough and wave crest respectively.*

Cancelling terms, we find that:

$$\frac{1}{2}(4cv) = 2gh$$

And since $h = \frac{vz}{c}$

Thus $cv = g\frac{vz}{c}$

Or $c^2 = gz$, so finally $c = \sqrt{gz}$ (**eq 2.26**) for shallow-water waves.

Thus c depends only on the depth of the water body, a very different result when compared with deep-water waves. If Z and h are both small compared with λ, this approach is suitable for long shallow waves, such as coastal tides with wavelengths that can be up to several hundred miles long. The main difference between deep- and shallow-water waves is that the circular orbital motion of the water particles has now been replaced by an almost completely horizontal longitudinal motion. No assumption has been made about the actual wave profile. It is seen that, provided the slope isn't too steep, the same argument may be applied to the velocity of transmission of any realistic surface or of an individual wave.

Example 2.5: A tidal shallow-water wave of period 12.4 hours exists in an estuary 45 m deep. What would the wavelength of the wave be (4 significant figures)?

The dispersion relationship is that:

$$\omega = \frac{2\pi}{T} = k\sqrt{\frac{gH}{k}} = \frac{2\pi}{\lambda}\sqrt{gH}$$

So rearranging for wavelength:

$$\lambda = \sqrt{gH}\,T = \sqrt{9.81 \times 45} \times 12.4 \times 3600 = 937.9 \ km$$

2.7.1 Dispersion

This is the gradual separation of wave types based on their relative wavelengths and speeds. Waves can also exhibit amplitude dispersion, where waves of larger amplitude have different phase speed from smaller amplitude waves. The shallower the water, the greater the interaction between a wave and the sea bottom and this can alter the wave properties, eventually causing the wave to collapse. Wave speed decreases as depth decreases. Troughs become flatter and the wave profile becomes very asymmetric. Wavelength decreases as depth decreases. Wave height increases as depth decreases while the period remains unchanged.

2.7.2 General summary equation covering both deep- and shallow-water wave types

So in shallow water $c = \sqrt{gz}$ but in deep water $c = \sqrt{\left(\frac{g\lambda}{2\pi}\right)}$. To establish a general speed equation requires an equation of the form:

$c = \sqrt{(\frac{g}{k}tan(kd))}$ since $tan(kd)$ for small kd tends to kd, thus $c = \sqrt{\frac{g}{k}kd} = \sqrt{gd}$

as expected for shallow water, while for large d $tan(kd)$ tends to 1 with

$c = \sqrt{\frac{g}{k}1} = \sqrt{\frac{g}{k}}$ so $c = \sqrt{\frac{2\pi g}{\lambda}}$, satisfying both deep and shallow limiting cases.

2.7.3 Wave measurements and devices

Visible observations are generally not very reliable. Height observation provides at best an approximate value of displacement only.

Instruments for measuring waves include: buoys, sub-surface pressure sensors, and lasers. Buoys with accelerometers can measure vertical acceleration. A *wave staff* may be attached to a maritime platform to record water height, measured by recording changes in capacitance or resistance of the wave staff. Pressure sensors can also be mounted below the surface to measure wave height.

Lasers may be attached to platforms and point down to the water surface. Satellite-based remote sensors provide data, e.g. radar altimetry data on waves has been provided by ERS-1 and ERS-2 satellites.

Synthetic Aperture Radar (SAR) provides successive radar observations from high-altitude aircraft or along satellite tracks; 3-D views of ocean surface topography are also available from the TOPEX/Poseidon satellites, giving satellite maps. Fluctuations in these measured height result from changes in ocean circulation.

2.7.4 Useful satellite sensing missions

Using the Magic Seaweed (MSW) website, you can establish the approximate wavelength of waves at various ocean locations on Earth, e.g. in the English Channel and approaches (magicseaweed.com).

Moderate resolution multispectral data from the Terra Satellite and Aqua Satellite, from the MODIS sensor (Moderate Resolution Imaging Spectroradiometer) is readily available (aqua.nasa.gov).

Argo Buoy Track Network

Argo is a global array of more than 3000 free-drifting profiling floats that measure the temperature and salinity of the upper 2000 metres of the oceans. This allows for continuous monitoring of temperature, salinity and velocity of the upper oceans, with all data being relayed and made publicly available within hours after collection. (www.argo.ucsd.edu/Argo_data_and.html)

TOPEX/Poseidon, a joint mission between the US space agency NASA and the French space agency CNES, was the first satellite mission to map ocean surface topography and revolutionised oceanography by providing satellite ocean observations. Its follow-on mission, JASON-1, was launched in 2001 to continue measurement of sea surface topography. The two satellites, TOPEX/Poseidon and

Jason-1, flew in tandem for three years, providing double the sea surface coverage and allowing scientists to study smaller features than could be seen by one satellite. TOPEX/Poseidon's radar altimeter provided the first continuous global coverage of the surface topography of the oceans. From orbit 1330 km above Earth, TOPEX/Poseidon provided measurements of the surface height of 95 per cent of the ice-free ocean to an accuracy of 3.3 cm. The satellite's measurements of the hills and valleys of the sea surface led to a fundamental new understanding of ocean circulation and its effect on climate, sealevel.jpl.nasa.gov.

The record of global sea surface height begun by TOPEX/Poseidon and Jason-1 continues with the Ocean Surface Topography Mission on the Jason-2 satellite, which launched in June 2008. Planning for a Jason-3 mission is now underway. Details of many other data sets can be found at sos.noaa.gov/Datasets.

2.7.5 Short comment on seismic waves

An introduction to different types of waves in the marine environment would, in our view, be incomplete if it did not also introduce seismic waves. These are produced by earthquakes and can, of course, generate tsunami waves. Two different types of seismic waves are produced, *body waves* and *surface waves*. Body waves travel through the Earth itself, while surface waves travel along the surface of the Earth, or along layers of the Earth's crust.

Body waves travel through the Earth through the processes of *reflection* and *refraction* – which will be discussed in some detail shortly – off various layers inside the Earth. Body waves are further split into two groups, P (pressure) or longitudinal P-waves, and S (secondary) transverse S-waves. Consequently, P-waves transfer energy through a series of compressions and rarefactions as discussed earlier when considering sound in air and sound in water. P-waves travel faster through the Earth than S-waves (no vertical oscillation is required), reaching measuring instruments such as suitably placed seismographs first. In S-waves, motions are transverse, where motion of the particles is perpendicular to the flow of energy.

$$V_p = \sqrt{\frac{K + \frac{4}{3}G}{\rho}} \qquad V_s = \sqrt{\frac{G}{\rho}}$$

Here are *equations showing the relationship between wave velocities and K (bulk modulus), G (rigidity modulus) and ρ (density) of the earth.*

As these waves travel through the Earth, they pass through areas of varying density ρ, bulk K and rigidity G moduli. Since the velocity of waves depends on these factors, one can find how these quantities vary at different depths.

Not only do waves undergo change in velocity when they encounter a boundary but they also reflect and refract in accordance with Snell's law. When P-waves refract off a boundary, some compression energy is changed into *shear* propagation in the secondary medium. This alteration from one type of wave to another is also possible from S to P waves. Seismology has found that there is an S-wave shadow on the opposite region of the Earth when an earthquake takes place due to the fact that shear waves cannot propagate through liquid, and so it was inferred that the outer core was liquid while the inner core is solid. There is also a P-wave shadow that appears at arc distances of about 103 degrees and 143 degrees corresponding to the P-waves approaching the core/mantle boundary. Although P waves penetrate the outer core boundary, if they arrive at the *critical angle* (θ_c) they can travel along the interface between the two media. When they enter the outer core they are strongly refracted and slow down due to the higher density of the outer core compared with the mantle.

2.8 Self-assessment questions

After studying this chapter, you should be able to answer the following questions:

1.   Show that the equation $y(t) = A\cos(\omega t) + B\sin(\omega t)$ will lead to the equation: $a(t) = -\omega^2 y(t)$ when velocity and acceleration are considered.

2.   If $y(t) = A\cos(\omega t) + B\sin(\omega t)$, find by differentiation the maximum and minimum values of $y(t)$.

3.   If $y(t) = A\cos(\omega t) + B\sin(\omega t)$ and the velocity is 3 ms^{-1} and the acceleration 6 ms^{-2}, find the general expression for the time t at which $v(t) = \dfrac{dy(t)}{dt} = 3 \; ms^{-1}$ and acceleration $a(t) = \dfrac{dv(t)}{dt} = 6 \; ms^{-1}$.

4.   For a pendulum where $\omega = \sqrt{\dfrac{g}{l}}$ and l is the length of the pendulum if f = 2 Hz and $\omega = 2\pi f$, find the length of the pendulum (2 decimal places).

5.   For waves in deep water, find the phase velocity for 50 m wavelength gravity waves.

6.   For shallow-water waves of wavelength 10 m, find the phase velocity.

7.   What is the ratio of deep- to shallow-water phase velocity?

8.   Deep-water wave amplitude is seen to decrease with depth according to the equation $A = A_0 \exp\left(-\dfrac{2\pi z}{\lambda}\right)$, where $A_0 = \dfrac{\lambda}{2\pi}$. Find the value A when $\lambda = 8\ m$ and $z = 4$ m (2 decimal places).

9.   Find the gradient of the curve (dA/dz) in Q8.

10.   Using the expression: $\dfrac{p}{\rho} + \dfrac{v^2}{2} + gH = constant$, find the final pressure if the final velocity is twice the initial velocity and the final height is half the initial height.

REFERENCES

[2.1] *Marine Physics*, RE Craig (Academic Press, London & New York, 1973, ISBN 978–0-1219–5050–7).

[2.2] 'Google Earth Physics', CE Aguiar and AR Souza *Physics Education* Vol.44 No.6 (2009), pp. 624–626 (www.if.ufrj.br/~carlos/artigos/PhysEd2009_GoogleEarth.pdf)

3

Electromagnetic waves

'The only thing worse than being blind is having sight and no vision.' Helen Keller

3.1 Background to electromagnetic wave motion

On a sunny day we can directly sense the light and heat travelling to the Earth's surface from the sun. Most of us are also aware that there are other invisible X-rays travelling from space to Earth, as well as radio waves travelling out into space from Earth to control space probes in deep space and those around distant planets and asteroids (such as the landing of Philae on the Comet Rosetta in November 2014, or robotic rovers on planets like the Mars Sojourner and Truth, which themselves can transmit radar, video and other information back to Earth again). All such waves can travel through the vacuum of space where there are no particles to support the wave propagation mechanisms we have just discussed for longitudinal and mechanical transverse waves in Chapter 2. It was this problem that James Clerk Maxwell sought to solve.

Part of Maxwell's understanding and approach to the solution of the electromagnetic wave problem was in considering the properties of both longitudinal waves (such as sound) and transverse waves (such as the waves set up in bridges and seismic waves), both types of wave that had been relatively well studied for over 150 years. Maxwell understood that longitudinal waves had only one fundamental direction: air molecules vibrated about their equilibrium position in the same direction as the propagation direction of the energy or communications information transmission. Such a wave could be considered as a one-dimensional wave with one degree of freedom or movement. A transverse wave, as shown in figure 1.2, has two degrees of movement – the movement of the connected particles is in a separate direction at 90 degrees to the direction of wave propagation – and could be considered as a two-dimensional wave.

Furthermore, it is possible in principle to demonstrate (and this can be achieved with a slinky spring) the excitation of a transverse motion in the direction coming out of the paper (sideways) towards you, as well as the one shown vertically in figure 1.2 with the propagation travelling in the plane of the paper as before. Thus there are two possible sinusoidal *modes* at 90 degrees to each other that could represent two different quantities at 90 degrees to each other as well. Consequently, the possibility existed in Maxwell's mind of two different *kinds* of transverse fields linked together, with one kind reaching a maximum at the same time as the other kind reached a minimum in field strength, transferring energy from one field to the other over time like a roller-coaster ride transferring potential energy to kinetic energy and then back again. The two connected, mutually dependent fields in the case of electromagnetic waves are somewhat unsurprisingly a time-varying electric field and a time-varying magnetic field, both having a sinusoidal variation.

It is observed experimentally that electric charges have electric fields in the space surrounding them. If the electric charges oscillate or vibrate then so will the electric fields. Moving charges, like those found in the mast of a radio antenna system, will also produce time-varying magnetic fields so an oscillating electric charge will produce an oscillating or time-varying electric field and an oscillating magnetic field, which are inextricably linked, creating a disturbance known simply as an *electromagnetic wave*. It can be shown mathematically that the electric field will always be at right angles to the magnetic field and both of these time-varying fields are perpendicular to the direction of the wave propagation. Since time-varying and static electric and magnetic fields can all exist in vacuum, electromagnetic waves can also propagate through a vacuum. An electromagnetic wave is portrayed in figure 3.1. It should be noted that the magnitude of the electric field is 300 million times larger than the magnitude of the magnetic field, which is sometimes called the magnetic induction (since $c = E/B$). And as these two different vector quantities are at 90 degrees to each other, they do not cancel out even when one is positive while the other is 180 degrees out of phase with it.

The full derivation of Maxwell's equations lies outside the scope of this book, requiring as it does vector notation and more complex operator functions. A full discussion of the derivation of Maxwell's important equations is given elsewhere [3.1].

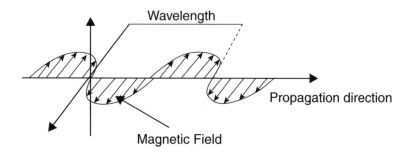

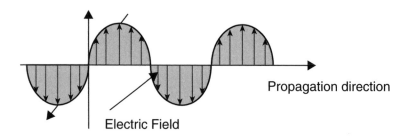

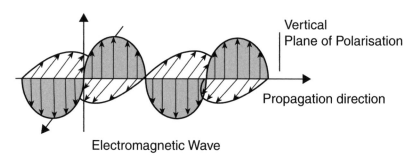

Figure 3.1: *Representation of an electromagnetic wave showing the orientation of both the electric and magnetic fields.*

3.2 Electromagnetic spectrum

3.2.1 Wave energy

As with all types of wave motion, we can specify the wave amplitude, the wavelength, the frequency and the speed of the waves: X-rays, γ-rays, ultraviolet radiation, infrared radiation, visible light, microwaves, radar and radio waves are all examples of electromagnetic waves. We will concentrate here on bands outside of the visible spectrum, as the visible spectrum – red to violet colours – is extensively documented in existing literature.

In principle, there are no fundamental differences between an ultraviolet wave and a microwave, except that these waves will interact differently with the matter they pass through depending upon the specific bond vibrations present in those materials.

They will differ in their frequencies and consequently their energies according to equation 1.3. Arranging the waves in order of increasing frequency (or decreasing wavelength) yields the *electromagnetic spectrum* as shown in figure 3.2 (see plate section).

With electromagnetic waves, the energy of a single photon or 'wave-packet' of radiation of a set frequency given by f:

$$E = hf \qquad \text{(eq 3.1)}$$

where h is Planck's constant having a value of 6.626068×10^{-34} m² kg s⁻¹, and f is the frequency as discussed previously. Using the result of equation 1.3 and substituting for the value of the frequency gives the result:

$$E = hc/\lambda \qquad \text{(eq 3.2)}$$

Example 3.1 If a wave has a wavelength of 2×10^{-6} m, what is the energy of the wave in both joules (J) and in electronvolts (eV) (both to 3 significant figures)?
$E = hc/\lambda = 6.62 \times 10^{-34} \times 3 \times 10^{8}/(2 \times 10^{-6}) = 9.93 \times 10^{-20}$ J
$1\,eV = 1.6 \times 10^{-19}$ J so measured in electronvolts:
$$= \frac{9.93 \times 10^{-20}}{1.6 \times 10^{-19}} = 0.621 \ eV$$

3.2.2 Radio waves

Radio waves are divided into various bands. Long wave radio, for example, can travel around the world and is still used for long-distance telegraphy. *Medium Frequency* (MF) waves are also used for night-time sky wave transmissions (up to 200 miles by day and 500 miles at night), relying upon *Total Internal Reflection* (TIR) from the charged ionospheric layers for their propagation. *High Frequency* (HF) waves are used for daytime sky wave broadcasting, operating over a higher frequency range and over a wider bandwidth (range of frequencies) than MF broadcasts. All sky wave transmissions are strongly subject to changes in the ionosphere from which they are refracted and through which they travel. VHF waves and frequencies above are not sufficiently strongly refracted by the Earth's atmosphere and ionosphere to be reflected back to the Earth, so they extend only to the horizon and a little beyond (typically 4/3 optical horizon approximately), passing through the ionosphere and then into space. These VHF waves are typically used for *Frequency Modulated* (FM) high quality radio broadcasts, which have wide signal bandwidth.

3.2.3 Microwaves

Microwaves is a very general description used to describe a part of the electromagnetic radiation with wavelengths ranging between 1 mm and about 50 cm. Microwaves include: UHF, EHF and the whole SHF bands (in terms of frequency 3 to 30 GHz, or in terms of wavelength 10 to 1 cm), (figure 3.2). Microwaves are used in many household items today and in radar systems on air, sea, space and land platforms. Long range systems operate below 4 GHz, where transmission loss is lowest. Short range systems operate up to 20 GHz ('J-band' 10–20 GHz).

Microwaves are generally 'small' compared with waves used in typical radio broadcasts, i.e. they have shorter wavelengths than radio wavelengths. Beginning at approximately 40 GHz, the atmosphere becomes less transparent to microwaves due to lower frequency absorption from atmospheric water vapour and at higher frequencies from atmospheric oxygen. Above 100 GHz, the absorption of electromagnetic radiation by Earth's atmosphere is so great that it is in effect opaque, until the atmosphere becomes transparent again in the so-called Infra Red (IR) part of the spectrum and in the optical part of the spectrum. However, the highest frequencies below 100 GHz give the greatest opportunity for high resolution imaging and military applications and are thus very short range.

Maritime users make extensive use of microwaves in the application of RADAR (Radio Aid for Detection and Ranging). Radar is an active method for obtaining the range of a contact. Radar transmits short pulses of electromagnetic energy in the microwave region and measures the range of the contact from the elapsed time between transmission to reception of the much weaker echo pulse. Most radar systems only give a measure of the strength of the target return, but the latest experimental systems can obtain an image of a ship or aircraft, even from beyond the visible horizon. Further discussion of radar is found in chapter 2 of Reeds Marine Engineering and Technology Series, Volume 15: *Electronics, Navigation Aids and Radar Theory for Electrotechnology Officers* (ISBN 978–1–4081–7609–2).

3.2.4 Infrared

The short *Near Infra Red* (NIR) wavelengths, between 0.7 and 3 μm (μm pronounced 'mu' or *microns*, i.e. 0.7 to 3 millionths of a metre), are used by Image Intensifiers (II[2]) and are commonly referred to as *night vision devices*). They are often seen on wildlife, police and military television programmes and can be combined with low light level Closed Circuit TeleVision (CCTV) systems (figure 3.3, below, and 3.4 – see plate section). These wavelengths are invisible to human eyesight but sensitive devices can make use of the large amount of invisible ambient starlight and moonlight, as well as amplifying any low level visible light present at night, especially in well-lit urban areas. NIR also penetrates cloud and fog a little more readily (less scattering) than visible light but still represents an operational flight risk as aviators generally use image intensifiers during night-time flying operations, and flying into cloud, fog, sandstorm or smoke will thus significantly reduce vision capabilities.

Figure 3.3: *Benefits of NIR imagery for zoological applications [3.2–3.3].*

Figure 3.3 shows a daytime NIR image of a zebra taken at Paignton Zoo, England. Notice the high NIR reflectance of vegetation (grass) and the extension of zebra stripe dye pigmentation well into the NIR region of the spectrum. Modern military camouflage must also account for the NIR properties of vegetation in order to blend into the natural background. A typical image intensifier will have a green phosphor display as the human eye is more sensitive to green than any other colour. However, image intensifiers respond to both visible and NIR radiation and both of these wavelength ranges will be displayed together as green (figure 3.4) unless additional filtering is introduced. If camouflage material does not have a high NIR reflectance, it will not match the background vegetation well. Note the high NIR reflectance of vegetation again taken here in the daytime using an NIR filter, which demonstrates that these devices can be used not only at night but during the daytime as well. In fact, daytime NIR image intensifiers can achieve better long range imaging capability than an ordinary visible camera, due to the fact that NIR has a reduced scattering and enhanced transmission through haze than visible light. Figure 3.4 (see plate section) shows a woodland image and path with high NIR reflectance.

Infrared waves, beyond the end of the red visible light spectrum, are shorter in wavelength than radio waves and are radiated by hot bodies, such as the sun, aircraft engines or a ship's hot funnels. Infrared waves are invisible to the naked eye, but produce the sensation of heat on human skin. Infrared radiation was first discovered by the celebrated astronomer Sir Frederick William Herschel (1738–1822). Although born in Hanover, Germany, Herschel emigrated to Britain at the age of 19. He made many important contributions to science and physics while living in Bath, England, and is best remembered for his discovery of Uranus and two of its moons, Titania and Oberon.

In February 1800, Herschel was testing solar filters to observe sun spots. When using a red filter he found a significant amount of heat was produced, and discovered infrared radiation in sunlight by passing it through a prism on to a series of thermometers, including one thermometer just beyond the red end of the visible spectrum. The thermometer was meant to be a control to measure the room's ambient air temperature, so he was surprised to discover it recorded a higher temperature than all the other thermometers in the visible spectrum. Further experiments led Herschel to conclude that there was an invisible form of light that extended beyond the red end of the visible spectrum.

Thermal imagery provides an extremely useful maritime tool for sea Search And Rescue (SAR) operations, surveillance, sea safety and vessel identification (figure 3.3) among other applications such as predictive maintenance and engine monitoring [3.4–3.5]. Modern commercial thermal imagers are very good tools for the maritime firefighter properly trained in their use, and can help to successfully contain fires. In the marine environment, a *Thermal Imaging Camera* (TIC) is an essential device today; unlike an image intensifier, it can see through thick black smoke. These are discussed in more detail in *Reeds Introductions: Essential Sensing and Telecommunications for Maritime Applications.*

In figure 3.5 (see plate section), a vessel is observed with a FLIR™ Systems E320 heat camera with a screen palette chosen to display images in a range of false colours to aid human observers. Cameras may be radiometrically calibrated to provide accurate temperature readings as well, and have proven valuable in wildlife research [3.6–3.7].

The *Middle Infra Red* (MIR), from about 3 to 6 microns wavelength, is commonly associated with very hot civilian heat sensing and military operation of heat-seeking missiles such as the USAF AIM-9 Sidewinder, which can passively seek the heat emissions of hot objects such as aircraft exhausts or ships outlined against a cold sea and cold sky [3.8].

Far Infra Red (FIR) thermal imaging cameras, operating from 6 to 15 microns, cover the temperature range associated with both human and wildlife emissions. SAR makes extensive use of thermal imaging, and can provide additional image identification at night in the absence of visible light or under conditions of mist and fog [3.4]. Radar may detect a target but a TIC provides a picture of the target, which can be extremely useful. However, the picture is not the same as we would see in high resolution visible and NIR but is a picture of the longer wavelengths emitted and reflected from an object, so a TIC user will need suitable training to properly make sense of the imagery they are observing.

3.2.5 Ultraviolet radiation

Ultraviolet (UV) radiation, shorter than the visible wavelengths such as infrared, is also invisible and is present in solar radiation. Ultraviolet radiation is responsible for human sunburns, but is also used positively by the body to produce vitamin D. Fortunately, high solar output levels of UV radiation are absorbed by ozone (O_3) in the upper atmosphere, but high levels of UV exposure can be encountered at high altitude.

Ultraviolet, like infrared, can be used to affect photographic films and various digital sensors; it can also help detect certain substances through its ability to fluorescence.

Certain materials, when exposed to ultraviolet radiation, are excited momentarily to a higher energy state, and upon relaxation drop back to their normal or 'ground' state, emitting the difference in energy at visible longer wavelengths. For example, specific minerals will emit particular identifiable spectral fluorescence spectra. Clothes achieve this effect using fluorescent substances that respond to ultraviolet waves present in ordinary sunlight or under illumination at music events. Oil can be detected on seawater by the use of UV fluorescence, when a powerful visible laser source illuminates the sea. In the absence of oil there is no significant backscattered fluorescence but with oil present there is usually a large backscattered visible fluorescence. Recent planar optical waveguide captured fluorescence from a small sample (1 cc) of Castrol GTX subjected to UV illumination is shown in figure 3.6. Specific fluorescence spectra can act as a unique 'fingerprint' of particular sources of oil and for other substances.

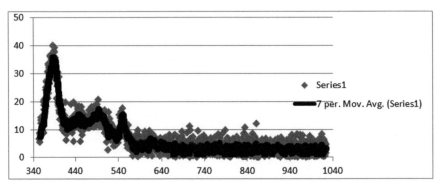

Figure 3.6: *Castrol GTX waveguide captured fluorescence spectra. Photon counts against wavelength (nm).© CR Lavers.*

3.2.6 X-rays

X-rays are yet another kind of invisible electromagnetic wave or radiation of even higher energy than UV, sometimes called Röntgen rays after the German scientist Wilhelm Conrad Röntgen (1845–1923), who discovered them in November 1895. His work was subsequently developed further by Marie Curie (1867–1934), who was born in Warsaw (Poland) as Marya Sklowdowska. In Paris, she graduated in physics and mathematics and married French physicist Pierre Curie. She studied radioactivity under Henri Becquerel (1852–1908), discovering both polonium and uranium, and was twice awarded the Nobel Prize: in 1903 for physics and in 1911 for chemistry. Her radiology equipment provided excellent services to thousands wounded in World War I and, with the help of her daughter Irène, Marie installed equipment in various field hospitals behind the Belgian frontline, mainly in 1914 and 1915. Marie herself died from the effects of radium in 1934.

Röntgen produced and detected electromagnetic radiation in the wavelength range today known as X-rays, an achievement that earned him the first Nobel Prize in Physics in 1901. X-rays' practical benefits arise from their ability to pass through considerable thicknesses of solid substances such as wood, brick or metal, which would stop ordinary light. X-rays also pass easily through the human body, with denser body materials such as bones absorbing more of the radiation than softer tissue. Modern medical digital X-ray machines convert the transmitted radiation into digital images of the interior of the human body. Dental radiographers can also use X-rays to examine the roots of teeth (figure 3.7).

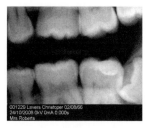

Figure 3.7: *Typical dental X-ray. Good shape or not?*

The shorter the wavelength of the X-rays, the greater the thickness of materials they will penetrate. Short wavelength X-rays are called 'hard' X-rays, and longer ones are called 'soft' X-rays. Hard X-rays are used to treat certain kinds of clinical diseases, such as cancer.

3.2.7 Gamma rays

Gamma rays, which are more penetrative than 'hard' X-rays, have higher frequency and even more energy. They are also used for treatment of sicknesses and the examination of solid objects, as well as being produced through nuclear reactions and detonations. Paul Villard (1860–1934), a French physicist, is largely credited with the discovery of gamma rays. Most accepted sources put this discovery at about 1900. Villard recognised gamma rays as different from X-rays because the gamma rays had a significantly greater penetrating depth. In 1914, the New Zealand-born physicist Ernest Rutherford (1871–1937) – later 1st Baron Rutherford of Nelson – showed that they were a form of light with even shorter wavelength than X-rays.

3.3 Radioactive decay and half-thickness

The radioactive decay of nuclear sources can emit powerful X-rays, which are extremely dangerous and hazardous to humans. To shield or protect radiation workers from X-ray sources usually involves a thick layer of lead or an even thicker layer of concrete if we are considering a building or fixed structure. It is common to quote the *half thickness* of shielding material that reduces the radiation level to half of its original (unshielded) level. Now, the intensity of X-ray radiation passing through metal decays exponentially according to the law:

$$I_x = I_o e^{-ax}$$ **(eq 3.3)**

where I_x is the intensity at a distance x into the specified material. I_o is the initial unshielded intensity level; a is the radioactive decay constant for the material in m^{-1} and x is the distance measured. Note: The unit for a, the radioactive decay constant, is given in terms of m^{-1} considering the near parallel flux over a short thickness, although in fact radiation from a source will radiate out into free space as the inverse square of distance. In nature, it is found that many things grow and decay in an exponential way. Exponential growth and decay is discussed in Appendix D.

Example 3.2: Consider an X-ray source. What will be the thickness of absorber required for the radiation level to be reduced to ¼ of its initial value in terms of a, if $a = 2\ m^{-1}$ (3 decimal places)? What is the relationship between x and a?

Use the equation: $I_x = I_o e^{-ax}$

Rearranging:

$$\frac{I_x}{I_o} = e^{-ax}$$

$$\frac{1}{4} = e^{-ax}$$

Taking inverse logarithms: $-\dfrac{\ln\left(\frac{1}{4}\right)}{a} = x$

So for $a = 2\ m^{-1}$

$$x = -\frac{\ln\left(\frac{1}{4}\right)}{2} = 0.693\ m$$

x is inversely proportional to a.

Example 3.3: What will be the value of the attenuation coefficient if the layer thickness required to reduce the unshielded intensity to half its initial value is 3 cm (3 decimal places)?
Taking the same formula as in the previous example and substituting for the values given: $\dfrac{1}{2} = e^{-0.03a}$ and rearranging the formula for a and taking logs to base e.

$$a = -\frac{\ln\left(\frac{1}{2}\right)}{0.03} = 23.105\ m^{-1}$$

3.4 General electromagnetic wave properties

All these electromagnetic waves travel with the same speed in a vacuum (i.e. the speed of light, $c = 2.98 \times 10^8$ m s^{-1} which is usually approximated to 3×10^8 m s^{-1}). In a material medium the wave speed is reduced and the ability to travel through the material, and the speeds with which they do so, depend upon the particular material and the frequency of the electromagnetic radiation. With visible wavelengths, such as light passing through a glass prism, white light will be observed to disperse or separate out into individual colours or wavelength because different wavelengths 'see' or experience a different refractive index and hence travel at a slightly different wave speed. Hence the shorter blue wavelengths experience greater refraction or deviation from the original direction of the beam of light than the red. Because the electric field (and the magnetic field) oscillates (vibrates) at right angles to the direction of propagation, all electromagnetic waves are considered as transverse waves.

It needs to be emphasised that things do not always appear the same when viewed in different spectral bands of the electromagnetic spectrum. Consider a view of the Dartmouth Castle Tea Rooms, England. For over 600 years, Dartmouth Castle has stood in an imposing waterfront setting, guarding the narrow entrance to the Dart estuary and access to Britannia Royal Naval College. The gun tower is thought to be the first fortification in Britain to mount heavy cannon. The castle battlements, Castle Tearooms and St Petrox Church together form an impressive heritage site. The same view can be obtained from several consecutive images during daytime in the visible red (figure 3.8), NIR (figure 3.9) and FIR bands (figure 3.10) [3.9–3.10].

Figure 3.8: *Visible red band.*

Figure 3.9: *NIR band.*

Figure 3.10: *FIR thermal band.*

Figures 3.8 and 3.9, taken in the visible and NIR bands respectively with appropriate spectral filters, seem at first glance to be superficially similar, but on further inspection significant differences are observed. For example, vegetation has a high NIR reflectance so that grass and trees appear 'bright'. Some regions of vegetation and objects even swap contrast (dark to light and vice versa), on-going from the visible to the NIR band. NIR also penetrates vegetation to a greater depth than the visible wavelengths, improving sub-canopy penetration. The resolution or quality of the visible and NIR are both high, while the FIR's thermal band resolution (figure 3.10) is much poorer by comparison. This lack of thermal resolution is due to the limited number of pixels found in thermal imagery sensors, which are usually (320 × 240 pixels) giving some 76.8 thousand pixels, unlike a modern digital camera which can now routinely exceed 20 megapixels. However, the thermal emissions of objects and people in the scene are detectable even on the darkest night (figure 3.10), when unaided visible and NIR imaging is not possible. Hence the value of thermal imaging in a maritime environment, especially for navigational awareness and for SAR missions, which may be required during the hours of complete darkness.

3.5 Self-assessment questions

After studying this chapter, you should be able to answer the following questions.

1.   List the following in order of frequency, lowest first: 10^{10} GHz SHF, 10^{-10} X-rays, 10 cm *echo sounder waves*, 10 cm radar, 1500 m radio waves, and blue light (400 nm).

2.   The following list shows five different wave motions:

3 GHz radar

Ultraviolet light

3 MHz radio signal

15 kHz sonar

1500 Hz sound in air

a. Which has the shortest wavelength?
b. Which has a higher frequency than visible light?

3.   Place the following in order of frequency, starting with the shortest:

5 GHz radar, 3 micron infrared radiation, 20 kHz sonar, 4 kHz sound in air, blue light, and 2000 m radio.

4.   For a single photon of wavelength 600 nm, what is the energy of the radiation emitted (3 significant figures)?

5.   For the photon of question 4, what is the energy of the photon in electronvolts (eV) (3 decimal places)?

6.   Consider an X-ray source. What will be the thickness of absorber required for the radiation level to be reduced to $\frac{1}{7}$ of its initial value in terms of a if $a = 0.76$ m^{-1} (3 decimal places)?

7.   Using differentiation and the equation $I_x = I_o e^{-ax}$, show the rate of change of intensity with distance.

8.   If the rate of change of intensity at a distance x is 3 Wm^{-1} and the intensity at distance x is 12.4 W, what is the decay constant a in m^{-1} (2 decimal places)?

9.   State the different spectral bands and their likely sensing and communications applications.

10.   If the individual photon packet of an X-ray has a frequency of 2.4×10^{19} Hz and the overall intensity of the X-rays is recorded as 1.6×10^{-11} Wm^{-2} across an area of 1 square metre in 1 second, what will be the total number of X-ray photons produced with constant flux (to the nearest whole number)?

REFERENCES

[3.1] *Electromagnetic Fields and Waves*, P Lorrain and D Corson (W.H.Freeman and Company, New York, 1970, ISBN 978–0–7167–0331–0).

[3.2] 'Application of remote thermal imaging and night vision technology to improve endangered wildlife resource management with minimal animal distress and hazard to humans', CR Lavers, K Franks, M Floyd and A Plowman, *Journal of Physics: Conference Series 15 Sensors and their Applications XIII* (2005) pp. 207–212.

[3.3] 'Application of remote thermal imaging and night vision technology to improve endangered wildlife resource management with minimal animal distress and hazard to humans', CR Lavers, K Franks, M Floyd and A Plowman, Proceedings of the Remote Sensing and Photogrammetry Society Annual Conference, with the NERC Earth Observation Conference, 'Measuring, Mapping and Managing a Hazardous World' Portsmouth University (2005).

[3.4] 'The Technology and Applications of Thermal Imaging', CR Lavers, *Electronics and Beyond*, No. 126 (June 1998), pp. 30–35.

[3.5] 'FLIR for Safety', B Dagiliatis, *Defence Helicopter* (September 1996), pp.40–45.

[3.6] 'Non-destructive high-resolution thermal imaging techniques to evaluate wildlife and delicate biological samples', CR Lavers, P Franklin, P Franklin, A Plowman, G Sayers, J Bol, S Shepard and D Fields, *Journal of Physics: Conference Series 178, No.1, Sensors and Their Applications XV* (2009) (Proceedings *doi: 10.1088/1742–6596/178/1/012040* ISSN: 1742–6588).

[3.7] 'What Heat Can Reveal – using thermal imagery for wildlife research', CR Lavers, *The Wildlife Professional* (Winter 2009), pp. 66–68.

[3.8] '*IR SAMS down but not out*', P Donaldson, *Defence Helicopter* (December 2003/January 2004), pp.23–26.

[3.9] A composite analysis between thermal images, visible and Near Infra Red photography of the archaeological site at Dartmouth Castle, Devon, England', CR Lavers, M Al Qattan, T Mason, K Franks, M Garcia and M Floyd, Proceedings of the Remote Sensing and Photogrammetry Society Conference, Nottingham University (September 2003).

[3.10] 'Observer Based Comparative analysis between thermal, visible and Near Infra Red images recorded at Dartmouth Castle, England', CR Lavers, K Franks, M Garcia, M Floyd, T Mason and M Al Qattan, Proceedings of the Remote Sensing and Photogrammetry Society Conference, Aberdeen, Scotland (September 2004).

4

Wave Properties

'Aye, I suppose I could stay up that late.'
James Clerk Maxwell, *after being informed of a
compulsory 6am church service at Cambridge University.*

In this chapter we will look at some further wave properties that are important
when considering electromagnetic waves and their interactions in a real
environment, such as polarisation, reflection, refraction and Total Internal Reflection,
along with a number of their maritime applications. All of these properties are
applicable to acoustic sound waves in the above water environment and sonar
acoustic waves underwater, with the exception of polarisation, which is a transverse
wave phenomenon only – and it is with polarisation we will begin.

4.1 Polarisation

Polarisation occurs only with transverse wave motion. Polarisation is a term
describing the alignment of the oscillation direction. Figure 4.1 shows a single
electromagnetic wave generated by an oscillating charge. The electric field
oscillation in this case is in the vertical direction and is a vector quantity, having
both magnitude (size) and direction.

If all the electromagnetic waves – and there may be many hundreds of millions
of them in a beam from a source such as radar or a laser – have their electric field
oscillations all in the same direction, the beam is said to be *linearly polarised*.

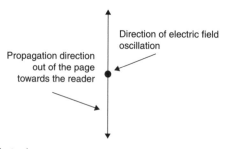

Figure 4.1: *Vertically polarised wave.*

With electromagnetic radiation, if the electric field oscillation is found to be perpendicular to the surface or ground, the beam is described as *vertically polarised* or transverse magnetic (as the magnetic field component is horizontal), whereas if the electric field is horizontal, the beam is described as *horizontally polarised* or transverse electric (as the electric field is now horizontal or across the surface). Sometimes, vertical polarisation is referred to as *P polarisation* and, similarly, horizontal polarisation with electric polarisation is known as *S polarisation* (figure 4.2). (This is also worth considering in the light of seismic waves discussed in the previous chapter.)

It is also possible to produce waves where the electric field direction spirals continuously clockwise (or anticlockwise) about the direction of propagation during transmission. This is known respectively as right-handed or left-handed *circular polarisation*. Such beams are often used in rainfall radars and satellite communications, and will occur in cholesteric liquid crystal materials commonly found in watches, displays, computers and calculators. Liquid crystal layer structure can itself be investigated with the use of polarised light [4.1–4.3].

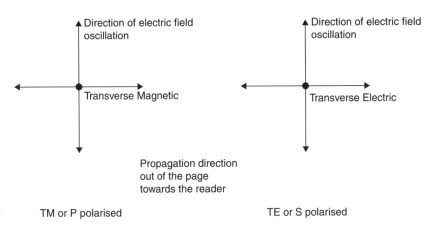

Figure 4.2: *Representation of TM and TE polarised waves.*

Usually, a wave polarised in one direction will interact with a surface and exit from the surface with the same polarisation, e.g. an S polarised wave incident on a flat reflecting surface will be reflected as S polarised radiation. However, some devices, such as Liquid Crystal Displays (LCD), rely upon S to P polarisation conversion to operate correctly, with the plane of polarisation being reflected within the liquid crystal layer. Further details about LCD can be found in chapter 14 of Reeds Marine Engineering and Technology Series, Volume 7: *Advanced Electrotechnology for Marine Engineers* (ISBN 978–1-4081–7603–0). If, however, the oscillation directions are randomly oriented with respect to the direction of propagation, the beam is described as *unpolarised*, and there is no preferred axis or direction (figure 4.3).

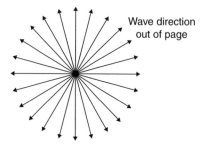

Figure 4.3: *Unpolarised wave.*

When an *unpolarised wave* reflects from a smooth horizontal surface, the reflections may contain a large proportion of waves with their electric field directions parallel to the reflecting surface. These reflections are called *partially linearly polarised* (or partially plane polarised).

Optical glare from a sea surface may be significantly reduced by viewing such a surface through a device that only allows vertically polarised light to be transmitted through it (e.g. a polaroid filter aligned to transmit vertically polarised light). The horizontal components, which make up most of the reflections, are therefore blocked and the glare minimised. Certain types of crystals and polymers orientated into sheets, such as Polaroid, have the property of 'filtering' light waves so that only the oscillations in one plane, either vertical or horizontal, will pass through. This can be demonstrated using a string and two slotted boards (figure 4.4). A transverse wave on a string will pass through a slot in the first board if the plane of polarisation is parallel to the slot. However, if the second board is rotated through 90 degrees, the wave motion cannot pass through because the oscillation of the string in the slot has been stopped. Thus if the slots are 'crossed', no vibrations will pass through; this is often referred to as *crossed polaroids*.

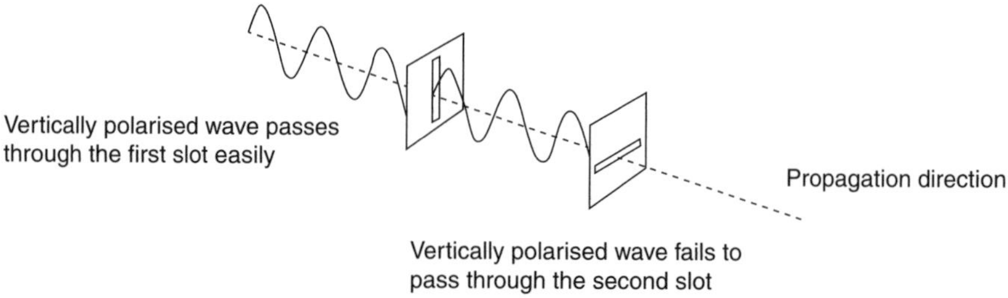

Vertically polarised wave passes
through the first slot easily

Propagation direction

Vertically polarised wave fails to
pass through the second slot

Figure 4.4: *Crossed polaroids.*

Figure 4.5 (left): *Image across the River Dart (polaroid parallel to a reflecting surface). Horizontal polarisation.*

Figure 4.6 (right): *Image across the River Dart (polaroid perpendicular to a reflecting surface). Vertical polarisation.*

However, the analogy used in figure 4.4 must be treated very carefully, as a polariser composed of a large number of vertical metal rods would be thought of as 'letting the vertically polarised waves through'. However, in reality, vertically polarised waves will excite electronic loss as electrons flow up and down the metal rod (or dipole) and are lost. Horizontally oriented waves do not excite resonant absorption in the rod, and pass through the screen – which at first thoughts, and in light of figure 4.4, used to explain the concept with a string, would seem counter-intuitive.

Unpolarised and polarised reflections from a sea surface are seen in figures 4.5 and 4.6 respectively. A surface tends to absorb light incident upon it strongly in the vertical orientation, but much less so in the horizontal orientation. In fact, an angle exists in the vertical orientation for which the reflected light will be a minimum, the so-called Brewster angle. Sir David Brewster (1781–1868) was a Scottish physicist, mathematician, astronomer, inventor, writer, science historian and university principal. He is mostly remembered for his contributions in the field of optics, discovering the photoelastic effect that gave rise to the field of optical mineralogy, and is recognised as the inventor of the kaleidoscope. Brewster's angle (or polarisation angle), named in honour of his explanation of this subject, is the angle of incidence at which light with a particular polarisation is perfectly transmitted through a transparent dielectric surface, with no reflection. When unpolarised light is incident at this angle, light that is reflected from the surface is thus perfectly polarised.

Note the reduction in *glare* from a fairly diffuse surface by comparison of figure 4.6 with figure 4.5, even in very poor quality weather conditions. To the crew on the bridge of a ship in bright sunlit conditions, the reduction in glare made by Polaroid glasses or polarising glass in the bridge windows is substantial, significantly improving what the watch-keeper or navigator can see. With the reduction in glare, the desaturating effect also means we can now observe enhanced detail in the previously otherwise saturated regions of figure 4.5.

When long-range, ship-borne early warning radars are used, interference between the direct beam to the target and the reflected beam from the sea surface can cause variations in the radar coverage above the sea surface, such that a target may not be detected. Using vertically polarised beams from the transmitter can minimise the intensity of the reflected beam and thus reduce the interference effects (to be discussed shortly) as again the horizontal components, which make up most of the sea surface reflections, are blocked and the problem of radar beam interference is minimised.

Clearly, the presence of vertically and horizontally polarised transmissions can provide problems of detection in communications. An aerial (or antenna) system with horizontal orientation will not detect vertically polarised waves, and vice versa. The alignment of the antenna will define the direction of the current and electron flow and hence the polarisation direction of the electric field.

Hence, to ensure signal detection under all possible circumstances, aerial inclination is best suited at 45 degrees to the vertical (or horizontal) to ensure both detection of horizontal and vertical polarised communication transmissions equally. This will, of course, mean that the detected strength of both polarisations will not be at its optimum (electric field is a vector quantity, having both magnitude and direction) – but at least the signal should now be detected. If a light source is aligned with its electric field E_0 at angle θ to the polarising x-axis (figure 4.7), the component of the electric field along the x-axis is $E_0 \cos \theta$. As the intensity of the light is given by the square of the electric field, I is proportional to $E_0{}^2\cos^2\theta$ or $I = I_0{}^2\cos^2\theta$ (**eq 4.1**). Clearly, as θ – the angle between the two polarisers – changes, the intensity of the transmitted intensity will vary also. This is often referred to as Malus's law.

$E_y = E_0 \sin \theta$

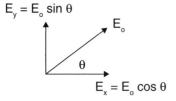

$E_x = E_0 \cos \theta$

Figure 4.7: *Components of electric field in the vertical and horizontal direction.*

4.2 Reflection of waves

When a wave strikes a smooth or flat surface, the incident and reflected waves travel in directions that are at equal but opposite angles to the normal from the reflecting surface, and are both in the same plane. The normal is the dashed line in figure 4.8, at 90 degrees to the reflecting surface. The angles are called the angle of incidence θ_i and the angle of reflection θ_r respectively. This is known as *regular reflection*, as shown in figure 4.8.

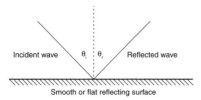

Figure 4.8: *Regular reflection, or reflection from a smooth surface.*

For many parallel waves incident from the same source at distance, the flat surface will result in a series of parallel waves, leaving the surface at the same angle of reflection (figure 4.9).

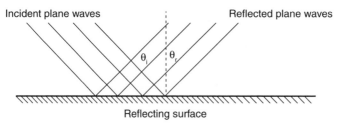

Incident plane waves   Reflected plane waves

θ_i θ_r

Reflecting surface

Figure 4.9: *Regular reflection of multiple plane waves from a smooth surface.*

If, however, the surface is rough and irregular, then the reflections will take place in a variety of directions and result in *diffuse reflection*. Small particles in the atmosphere or in the sea can cause multiple reflections and this is known as *scattering* (figure 4.10).

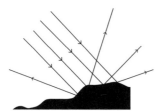

Figure 4.10: *Diffuse reflection.*

Waves incident at 90 degrees to a flat surface will be reflected strongly back directly along the normal towards their source, creating a strong reflection. Radar makes use of this principle so a platform at sea (or on land or in air or space) can obtain both target detection and a calculation of target range. To avoid this, naval warships are often designed with different parts of their superstructure at different angles of inclination so the ship never generates a particularly large reflected signal. A typical angle of tilt in the superstructure of, say, 5–10 degrees away from the vertical would result in a total deviated beam of 10–20 degrees correspondingly away from the probing radar system and hence is likely to generate a very weak reflected signal level. This is one of the basic underlying design features of modern stealth naval warships [4.4–4.6]. Further details about stealth warship design may be found in a good introduction to the topic [4.7].

This strongly angular dependent reflection condition phenomenon can in fact be observed with all electromagnetic waves and can be seen optically by comparison of figures 4.11 and 4.12 respectively (see plate section for both). In figure 4.11 we see the last dying rays of the sun before sunset generating a very strong, dazzling reflected signal back to the observer across the River Dart, Dartmouth, England, from Kingswear. Deviation of much less than 1 degree in angle will significantly reduce the reflected glare from the perspective of the observer (figure 4.12). Additionally, this strong solar reflection, or glint, is to be avoided with all use of optical instruments, as strong reflections from binoculars, etc., will generate tell-tale reflections that will reveal mariners involved in otherwise covert operations.

To avoid such reflections, single and multilayer anti-reflection coatings are used, often incorporating thin film layers of materials such as magnesium fluoride, an inorganic compound with the chemical formula MgF_2. The compound is a white crystalline salt and transparent over a wide range of wavelengths. Magnesium fluoride coatings are often used in terrestrial optics and in space telescopes. Due to its having a refractive index of 1.37–1.38, a single thin layer of MgF_2 is commonly used on the surfaces of optical elements as an inexpensive anti-reflection coating. Compound structures incorporating repeated layers of magnesium fluoride (low index layer) and a high refractive index layer can produce tailored anti-reflection coatings. The theoretical reflectivity as a function of incident angle for a prism of high index flint glass is shown in figure 4.13. The same prism is now coated with a thin layer (200 nm) of magnesium fluoride n = 1.378, which yields a marked reduction in reflectivity over a broad range of incident angles (figure 4.14).

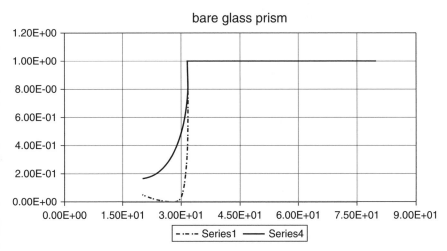

Figure 4.13: *Theoretical reflectivity of a bare glass prism with refractive index n = 1.678 (extra dense flint). Series 4 Vertically polarised wave. Series 1 Horizontally polarised wave.*

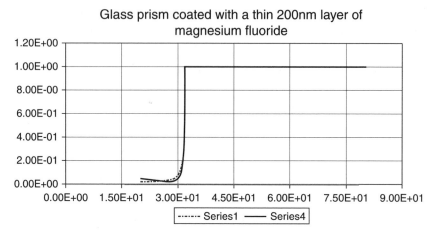

Figure 4.14: *Theoretical reflectivity of a glass prism with refractive index = 1.678 (extra dense flint) coated with a single low refractive index layer of magnesium fluoride n = 1.3678 and a high refractive index flint prism. Series 4 Vertically polarised wave. Series 1 Horizontally polarised wave.*

In radar, a similar reflection problem to that encountered for light directly reflected from the windows observed in figure 4.11 occurs at sea from two (dihedral) or three (trihedral) right-angled surfaces meeting together (figure 4.15).

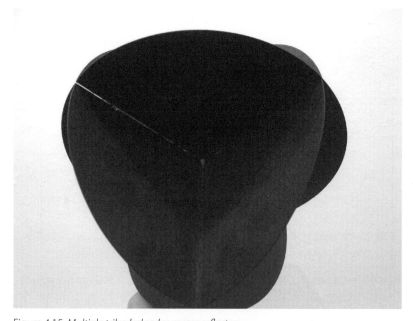

Figure 4.15: *Multiple trihedral radar corner reflector.*

These geometries are such to ensure very strong reflected signals back towards the radar system at all times. These so-called corner reflectors are ideal for merchant ships and yachts wishing to be seen passing through busy shipping lanes, but are not so ideal from the perspective of a naval warship. Ship design incorporates built-in deviations from the normal 90 degree angles, and older vessels can have plates suitably welded into place to destroy the 90 degree angle, while suitably deployed naval decoys (off board) can happily incorporate as many dihedral and trihedral surfaces as we might like [4.8].

4.3 Refraction of waves

4.3.1 Refraction of waves between different media

Refraction is the change in direction of a wave due to a change in its speed. This process is commonly observed when a wave passes from one medium to another at any angle other than along the normal. Refraction of light is the most commonly observed effect, but all electromagnetic waves will also undergo the same phenomenon, and in fact all types of waves can refract when they interact with a different medium, such as when sonar waves pass from one 'layer' into another, or when water waves move into water of a different depth.

Refraction is described by Snell's law, named after Willebrord Snellius (1580–1626), a Dutch astronomer and mathematician who is largely credited with the discovery of the calculation of light refraction. However, this phenomenon is known to have been investigated much earlier by the ancient Greek Ptolemy and by other later scientists such as Witelo and Alhazen of North Africa, who himself wrote several commentaries on the works of Ptolemy, Aristotle and Euclid. Snell's law states that the angle of incidence θ_i is related to the transmitted angle (angle of refraction) θ_t by the following equation:

$$N_i \sin \theta_i = N_t \sin \theta_t = V_t / V_i \qquad \textbf{(eq 4.2)}$$

where N_i and N_t are the refractive indices within the respective media, and V_i and V_t the wave speeds. In other words, the optical path of $N \sin\theta$ is a minimal constant in each medium so the light will always take this minimum (constant) value. It should be noted that the refracted wave is also the transmitted wave into the second media and that in general there will also be a reflected wave present at all times. In the case of total internal reflection, no refracted wave is transmitted into the second media.

Example 4.1: Consider a wave incident at a boundary between air and water. The air has a refractive index of 1.0003, while the water has a refractive index of 1.333. The wave is incident at an angle of 30 degrees to the normal, see figure 4.16. Find the transmitted angle θ_t in the water (the angle away from the normal), and sin θ_t (both 3 decimal places).

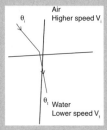

Figure 4.16: *Refraction from air into water.*

The refractive index of the air means that the speed of the light in air is ever so slightly less than that of light travelling in vacuum. From the formula above, the refracted (transmitted) angle can be found by using:

Ni sin θi = Nt sin θt

so sin θt = Ni sin θi / Nt

= 1.0003 × sin 30° = 0.375 (3 decimal places)

And θt = sin⁻¹ (0.375)

So θt = 22.037° (3 decimal places)

Consequently, when a wave travels from one region to another where the wave travels at a different speed, refraction will occur and the wave will deviate away from its original direction. If the speed decreases, the wave will deviate towards the normal to the boundary between the regions (figure 4.16). However, if the speed increases, for example, as it might going from glass into air, it will deviate away from the normal (figure 4.17). It is observed that the value of $Nx \sin \theta x$ is a constant for all x layers (or media).

Also as N = c/v or v = c/N

So Vwater = c/1.333 = c/(4/3) or Vwater = (3/4)c

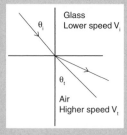

Figure 4.17: *Refraction from glass into air.*

4.3.2. A simple derivation of Snell's law

It has already been noted that light travels more slowly within a prism of glass than it would if a medium of air or vacuum were surrounding it. The consequent result of this difference in wave speed is observed in the bending of visible light rays. The refractive index of this experimental change is given by the ratio of the speed of light in vacuum to that in the second medium (in this case the speed of light in the glass), which can be measured by the degree of bending of the waves that occurs.

We can illustrate this phenomenon by a simple example. Consider a ray of light passing through air and entering a plane parallel layer of glass, as shown in figure 4.18. Now, Snell's law gives the relation between the angle of incidence (θ_i), the angle of refraction or transmission (θ_t), the refraction index of the glass slab (n_t) and the surrounding incident medium (n_i), such that:

$$n_i \sin \theta_i = n_t \sin \theta_t \qquad \textbf{(eq 4.2 repeated)}$$

In figure 4.18, the angle of incidence and the refraction angle are the angles between the light beams and the normal to the surface at the point where the beam crosses the interface.

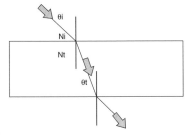

Figure 4.18: *A beam of light is bent as it enters a layer of glass.*

Now let's give a simple proof of Snell's law using figure 4.19 and Fermat's principle, which states that:

'A light ray will follow the path between two points that takes the shortest time.'

Considering a general case examining the waves in two adjacent media, media 1 and 2, Snell's law would be rephrased generally as:

$$n_1 \sin \theta_1 = n_2 \sin \theta_2 \qquad \textbf{(eq 4.3)}$$

where θ_1 is the angle of incidence, θ_2 is the refraction angle, the refractive index of the surrounding air is n_1 and the refraction index of the glass slab is n_2. The velocity of light in the two media is given by V_1 and V_2.

Consider the relationship velocity = distance/time taken and rearranging this for time, then: time = distance/velocity.

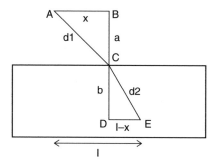

Figure 4.19: *Geometric proof of Snell's law.*

Now consider the time it takes to get from A to C and then to E in figure 4.19. t_1 is the time taken crossing medium 1 and t_2 the time taken crossing medium 2. The time is found by considering the two right-angled triangles labelled ABC and CDE and using Pythagoras' theorem, such that:

$$t = t_1 + t_2 = \frac{d_1}{V_1} + \frac{d_2}{V_2} = \sqrt{\frac{a^2 + x^2}{V_1}} + \sqrt{\frac{(l - x)^2 + b^2}{V_2}}$$

Now, according to Fermat's theorem we require the shortest or least time-consuming path for the light ray to take. To do this, we need to obtain the rate of change of time with respect to the x-coordinate, or the first time derivative with respect to x of the wave entering the glass layer. Setting this derivative or gradient to zero gives us a solution that will correspond to the *minimum* value of time and therefore the actual path the light wave will take. Derivatives and differentiation are discussed in Appendix C.

$$\frac{\partial t}{\partial x} = \frac{\frac{1}{2}(a^2 + x^2)^{-1/2}2x}{V_1} + \frac{\frac{1}{2}[(l - x)^2 + b^2]^{-\frac{1}{2}}2(l - x)(-1)}{V_2} = 0$$

And thus:

$$\frac{x(a^2 + x^2)^{-V_2}}{V_1} = \frac{(l - x)[(l - x)^2 + b^2]^{-\frac{1}{2}}}{V_2}$$

Rearranging:

$$\frac{x}{V_1\sqrt{a^2 + x^2}} = \frac{(l - x)}{V_2\sqrt{(l - x)^2 + b^2}} \qquad \textbf{(eq 4.4)}$$

Which dividing both sides through by x becomes:

$$\frac{1}{V_1\sqrt{\frac{a^2}{x^2} + 1}} = \frac{1}{V_2\sqrt{\frac{b^2}{(l - x)^2} + 1}} \qquad \textbf{(eq 4.5)}$$

Now from figure 4.19: $tan\ \theta_1 = \frac{x}{a}$ and $tan\ \theta_2 = \frac{(l - x)}{b}$

And by suitable algebraic manipulation of the 1/tangent, i.e. the cotangent, since:

$$cot\ \theta_1 = \frac{a}{x} \text{ and}$$

$$cot\ \theta_2 = \frac{b}{(l - x)}$$

Then equation 4.5 becomes:

$$\frac{x}{V_1\sqrt{cot^2\theta_1 + 1}} = \frac{(l - x)}{V_2\sqrt{cot^2\theta_2 + 1}}$$

And since:

$$sin\theta = \frac{1}{\sqrt{cot^2\theta + 1}}$$

So: $\dfrac{x\sin\theta_1}{V_1} = \dfrac{(l-x)\sin\theta_2}{V_2}$

Ignoring the geometric to scale-sized terms from the diagram (by setting $l = 2x$), and considering the relationship between angle and speed only, this is equivalent to:

$\dfrac{\sin\theta_1}{V_1} = \dfrac{\sin\theta_2}{V_2}$

And since the index of refraction is defined as $N = \dfrac{c}{V}$

Thus: $n_1 \sin\theta_1 = n_2 \sin\theta_2$

We can now revisit reflection in the same way. Given figure 4.8 and drawn for arbitrary θ_i and θ_r (figure 4.20), we can find the time taken for an arbitrary wave to travel to a reflected surface and then back to the same height above the reflecting surface.

$t_1 + t_2 = \dfrac{d_1}{V_1} + \dfrac{d_2}{V_1} = \dfrac{x}{\sin\theta_iV_1} + \dfrac{(l-x)}{\sin\theta_rV_1}$

And again taking the first derivative with respect to distance x:

$\dfrac{\partial t}{\partial x} = \dfrac{1}{\sin\theta_iV_1} + \dfrac{-1}{\sin\theta_rV_1} = 0$

Thus, rearranging yields the result: $\sin\theta_i = \sin\theta_r$ and so $\theta_i = \theta_r$

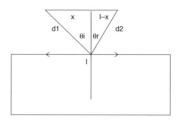

Figure 4.20: *Calculation of minimum path for reflection using differentiation.*

4.4 Total Internal Reflection (TIR) of a wave

When a wave travels from a slow speed region (having high refractive index) to a fast speed region (low refractive index), it will bend away from the normal, as described in figure 4.21. As the angle of incidence increases, eventually an angle will be reached where the angle of refraction will be exactly 90 degrees, i.e. the wave refracts along the boundary between the two regions and no transmitted or refracted wave will occur. The angle of incidence for which this occurs is known as the *critical angle* (θ_c). Any increase in incident angle beyond (θ_c) will result in *Total Internal Reflection* (TIR) of the wave.

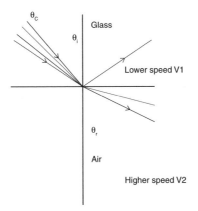

Figure 4.21: *Total Internal Reflection of a wave.*

TIR is easy to demonstrate with a glass prism, as illustrated in figure 4.22 (see plate section), with complete reflection observed above the critical angle, figure 4.23 [4.9].

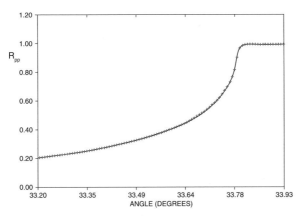

Figure 4.23: *Reflectivity for the ITO-coated prism shown in figure 4.22. Experimental data crosses, theoretical predictions the continuous line.*

Note the sharp rise in reflectivity approaching the critical angle in figure 4.23 near 33.78 degrees, and the use of Rpp (vertically polarised light incident) to denote p-polarised light entering and p-polarised light exiting the prism.

The importance of designating P (vertically polarised or TM) as opposed to S (horizontally polarised or TE) is not insignificant, as above the critical angle on a metal-coated prism it is also possible to excite a special surface travelling wave or Surface Plasmon Polariton (SPP), which can be used for a variety of sensing applications (figure 4.24). This SPP can be excited on a metal film attached directly to the prism in the so-called Kretschmann configuration [4.10] or separated a small distance from the prism, usually by an air gap, in the so-called Otto configuration [4.11].

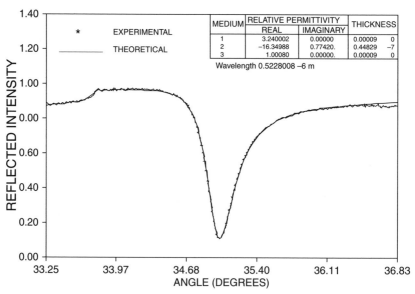

MEDIUM	RELATIVE PERMITTIVITY		THICKNESS	
	REAL	IMAGINARY		
1	3.240002	0.00000	0.00009	0
2	−16.34988	0.77420.	0.44829	−7
3	1.00080	0.00000.	0.00009	0

Wavelength 0.5228008 −6 m

Figure 4.24: *Silver-coated prism with resonant absorption 'dip' above the critical angle for P polarised light only. Experimental data crosses, theoretical predictions the continuous line [4.9].*

The mirage seen on a road on a hot summer's day is yet another example of total internal reflection. The air near the surface of the tarmac heats up during the day, expands and has a slightly lower refractive index than the surrounding cooler and denser air. Light approaching the surface is continuously refracted away from the normal until it is reflected back from the surface from above with an almost mirror-like quality. Figure 4.25 (see plate section), showing a 'mirage', was captured by the author; the reflection of the vehicle and its brake lights are even visible as it goes downhill as seen in the mirage, although the number plate cannot be read. The surface of the road appears as if wet, although it is of course completely dry, which is typical of the desert mirage.

Extensive use is made of this type of reflection phenomenon in many optical instruments and fibre optics, as well as being responsible for ducting (trapped guiding of the waves) in sonar beam propagation, radar transmission, and radio reflections from the ionosphere over various scales of size and wave types. Although modern submarine optics now make extensive use of optical fibres, older submarine telescope optics used TIR from prisms, as untreated metal-coated mirrors will degrade very quickly by the action of electrochemical etching in saline conditions.

4.4.1 Optical fibre

An *optical fibre* is a flexible, transparent fibre made of pure glass (silica) not much wider than a human hair, which acts as a waveguide, or *light pipe*, transmitting light between the ends of the fibre. Optical fibres are widely used in fibre-optic communications, permitting transmission over longer distances and at higher bandwidths (and data rates) than other forms of communication. Fibres are used instead of metal wires because signals travel along them with less attenuation (loss) and are more immune to electromagnetic interference. Fibres are used for illumination, and are wrapped in bundles so they can be used to carry images, allowing viewing in tight spaces, especially for medical endoscopic applications. Specially designed circular cross-section fibres are used for a variety of other applications, including sensors [4.12] and fibre lasers [4.13]. Flat planar waveguide sensor geometries are also possible [4.14–4.15].

Optical fibre consists of a transparent core surrounded by a transparent cladding material with a lower refractive index. Light is kept in the core by TIR, and causes the fibre to act as a waveguide. A fibre with a cut-away section is shown in figure 4.26 (see plate section). All fibre periscopes are possible, replacing the need for prisms. However, an electro-optical interface will prevent laser light passing through an all-optical path to enter the eye directly. The core/cladding structure will usually have a diameter of about 125 microns and will also include an outer plastic protective covering for a small bundle of many fibres, taking up significantly less space than conventional coaxial cable, and allowing millions of times more data to be transmitted.

4.4.2 Ionospheric refraction

The *ionosphere* is a part of the upper atmosphere, beyond the region of Earth's weather (the troposphere, up to 15 km), distinguished because low density of gas molecules in these regions (50–400 km altitude) can be ionised by incident solar radiation. Above the uniform mixing of the troposphere we find the stratosphere, where gas molecules tend to 'layer out' according to their molecular weight; oxygen being on average lower than the two isotopes of nitrogen layers because of its greater molecular weight.

The ionosphere has practical importance because it influences radio wave propagation around Earth. These molecules tend to become ionised by solar radiation as the sun rises and the refractive index of ionised air is found to be below that of non-ionised air. Consequently, refractive conditions can arise, which permit waves to undergo refraction and under certain conditions can undergo total internal reflection.

At certain times, some radio wave frequencies are totally internally reflected by 'transparent' charged gas layers in the ionosphere. Such waves are referred to as *sky waves*. Generally, sky wave transmission occupies the range 3–30 MHz during the day but a narrower band of 0.5–1.5 MHz at night, due to recombination of ions in the ionosphere.

Due to the transient nature of solar ionisation, ionospheric communications are not the most reliable mode of long haul communications. Several layers of charged particles allow varied working radio routes. An ionospheric layer is sketched in figure 4.27. Here, the ionisation of the layer reaches a maximum in the middle; before and after, this ionisation decreases and therefore so does the refractive power. If the incident wave is not sufficiently refracted by the time it gets to the middle, it will thus not be returned to Earth and escapes from the atmosphere as an escaping sky wave or space wave. Space waves are, of course, extremely important for satellite communications, without which such communication would not be possible using electromagnetic waves.

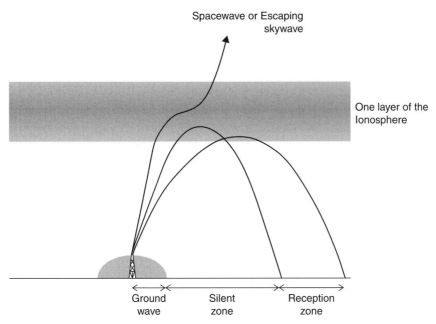

Figure 4.27: *One layer of the ionosphere totally internally reflecting waves from different incident angles.*

Guglielmo Marconi (1874–1937), whose team reputedly received the first transatlantic radio signal on 12 December 1901, in St. John's, Newfoundland, Canada, using a kite-supported antenna over 150 m long for reception, is considered the founder of modern radio communications. His transmitting station in Poldhu, Cornwall (Mullion Cove), UK, used a spark-gap transmitter to produce a signal with a frequency of about 500 kHz, 100 times more powerful than any radio signal previously produced.

The purpose of the spark gap is to initially present a high resistance in a circuit so that a capacitor charges. When the breakdown voltage of the gap is reached, air in the gap ionises, the resistance across the gap is much lower and a current pulse flows across the arc to the rest of the transmitter circuit. The gap is set so the discharge coincides with a maximum or near charge maximum in the first capacitor, C1. It is as if a high-speed switch is turned on at that moment to allow the capacitor to discharge its stored energy into the other circuit elements. This pulse rapidly moves back and forth between the capacitative and inductive elements, taking the form of a damped radio frequency oscillation, an alternating current with a voltage wave flowing out to the antenna.

Once perfected, this secretive experiment could have enabled Marconi's financial backers to get details of closing share prices in London to partners in New York before the American Exchange closed. However, the initial experiments were of a modest stature. The message received was 3 dits of the Morse code for the letter S. Transmission took place over a two-hour daytime period, understood likely to provide the best ionospheric conditions possible to guide electromagnetic waves across the Atlantic Ocean. To reach Newfoundland, the signal had to bounce off the ionosphere, due to the wall of water in the way created by the curvature of the Earth across the Atlantic. Of the numerous Morse Ss sent on that fateful day in 1901, only six were received. And yet this was a resounding success, demonstrating in principle that such wireless telegraphy messages could be sent using the ionosphere, without the need for wires.

In 1902, Oliver Heaviside (1850–1925) – an inventive mathematical genius who rose from humble origins to develop many mathematical ideas of his own, founded on Maxwell's earlier works – proposed the existence of the Kennelly–Heaviside layer that bears his name today. Heaviside's proposal included means by which radio signals could be transmitted around the Earth's curvature. Additional comments on the ionosphere are to be found in chapter 3 of Reeds Marine Engineering and Technology Series, Volume 15: *Electronics, Navigation Aids and Radar Theory for Electrotechnology Officers* (ISBN 978–1-4081–7609–2).

The term *ionosphere* was first introduced by the Scottish physicist Robert Watson-Watt in a letter written in 1926; it was published in 1969 in the journal *Nature*. Watson-Watt was responsible for developing the Chain Home Command Radar System for the RAF. The system arose out of an initial inquiry in 1934 by Sir Henry Tizard from the UK Air Ministry, who approached Watson-Watt to evaluate whether a radar 'death ray' was possible. Watson-Watt was able to show that this was not possible with pre-World War II technology but at low radio frequencies might be used to locate incoming Luftwaffe bombers, creating the first example of what we would now describe as *Network Centric Warfare* (NCW). The Chain Home Command System is largely credited with winning the Battle of Britain in 1940 due to its uncanny ability to vector a relatively small number of still functioning Royal Air Force fighter aircraft to intercept the numerous German bombers.

Edward Appleton (1892–1965) was awarded a Nobel Prize in 1947 for his confirmation in 1927 of the existence of the ionosphere. Lloyd Berkner (1905–1967) first measured the height and density of the ionosphere, which permitted the first complete theory of short wave radio propagation to be achieved. Further details about the early days of radar can be found in chapter 12 of Reeds Marine Engineering and Technology Series, Volume 14: *Stealth Warship Technology* [4.7].

4.5 Self-assessment questions

After studying this chapter, you should be able to answer the following questions.

1.   Explain the difference between polarised and unpolarised waves.

2.   Find by differentiation the minimum of the expression: $I = I_0{}^2\cos^2\theta$.

3.   Explain why long-range, ship-borne surveillance radars are vertically polarised.

4.   Explain the difference between regular and diffuse reflection and scattering.

5.   Describe how the effects of refraction and total internal reflection arise.

6.   Find the critical angle for a wave travelling from diamond (refractive index = 2.4) into air having refractive index = 1.0003. Find also the refracted wave angle to the normal in air for an incident wave from diamond at 15 degrees to the normal (2 decimal places).

7.   Use the following equation to calculate the missing variable in each situation: $\lambda = dx/L$ where λ is the wavelength of the light, x is the fringe spacing, d is the distance between slits, and L is the distance from the slits to the screen.

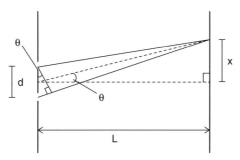

i) The wavelength of the light in the instance where x = 0.05 cm, d = 2.5 mm and L = 2.3 m (nearest whole nm).

ii)   What distance must the screen be from the slits for the fringe spacing to be twice the distance between the slits for blue light when x = 1 mm and L = 5 m?

8.   Consider a wave incident at a boundary between air and water. The air has a refractive index of 1.0003, while the water has a refractive index of 1.333. The wave is incident at an angle of 30 degrees to the normal (figure 4.16). Find the transmitted angle θ_t in the water, and $\sin \theta_t$ (both 3 decimal places).

9.   The figure below shows a light ray passing from air into an unknown substance x: The refractive index of air is 1.0006. Calculate the value of the index of refraction of the unknown substance x.

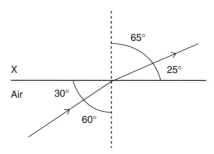

10.   Consider a wave incident at a boundary between water and glass. The air has a refractive index of 1.333, while the glass has a refractive index of 1.52. The wave is incident at an angle of 20 degrees to the normal. Find the transmitted angle θ_t in the water, and $\sin \theta_t$ (both 3 decimal places).

REFERENCES

[4.1] 'An examination of the optical dielectric tensor of a liquid crystal waveguide', CR Lavers and JR Sambles, *Ferroelectrics*, Vol.113 (1991), pp.339–351.

[4.2] 'The use of Mode Mixing to determine the Optical Dielectric Tensor in an FLC cell', CR Lavers and JR Sambles, *Liquid Crystals*, Vol.8, No.4 (1990), pp.577–585.

[4.3] 'Optical probing of thin liquid crystal layers using the prism-coupling technique', CR Lavers, *Liquid Crystals*, Vol.11, No.6 (1992), pp.819–832.

[4.4] 'Future Stealth Warship Design', CR Lavers, *Electronics World* (December 2007), pp.14–18.

[4.5] 'Invisibility Rules the Waves', CR Lavers, *Physics World* (March 2008), pp.21–25.

[4.6] 'Stealth Warship Design', CR Lavers, *Military Technology Journal*, Vol XXXII Issue 5 (June 2008), pp.42–51.

[4.7] Reeds Marine Engineering and Technology Series: Volume 14: *Stealth Warship Technology*, Christopher Lavers (Thomas Reed, 2012, ISBN 978–1-4081–7552–1).

[4.8] *Electronic Warfare and Radar Systems Engineering Handbook*, p. 67, NAWCWPNS TP 8347 w/Rev 2 (1 April 1999) and later changes PDF/Adobe Acrobat by EWC Desk Commander, Naval Air Forces Atlantic. www.microwaves101.com/encyclopedia/Navy%20handbook/EW_Radar_Handbook.pdf

[4.9] 'Determination of the optical dielectric constants, and deformational effects, after surface treatment, of a polyimide alignment layer used within a ferroelectric liquid crystal device system', CR Lavers, *Thin Solid Films*, Vol.289, No1–2 (1996), pp.133–139.

[4.10] 'Radiative decay of nonradiative surface plasmons excited by light', E Kretschmann and H Raether, *Zeitschrift für Naturforsch*, Volume 23 (1968), p. 2135.

[4.11] 'Excitation of nonradiative surface plasma waves in silver by the method of frustrated total reflection', A Otto, *Zeitschrift für Physik*, Volume 216, Issue 4 (1968), pp. 398–410.

[4.12] 'Fiber-Optic Sensing: A Historical Perspective', B Culshaw and A Kersey, Invited Paper, *Journal of Lightwave Technology*, Vol. 26 No. 9 (May 1, 2008), pp. 1064–1078.

[4.13] *Tunable Laser Applications*, FJ Duarte (Ed) 2nd Edition (CRC, New York, 2009, ISBN 978-1420060096).

[4.14] 'Electrochemically controlled optical waveguide sensors', CR Lavers, C Piraud, JS Wilkinson, M Brust, K O'Dwyer and DJ Schiffrin *Proceedings of Optical Fiber Sensors 9* (Florence, May 1993).

[4.15] 'A waveguide-coupled surface plasmon resonance sensor for an aqueous environment', CR Lavers and JS Wilkinson, *Sensors and Actuators* B, Vol 22. (1994), pp.75–81.

[4.16] 'Dielectric-Fibre Surface Waveguides for optical frequencies', KC Kao and GA Hockham, Proc. IEE, 113 (1966), p. 1151.

5

Loss Mechanisms

'For it is light that makes everything visible.'

Ephesians 5:14 New Testament (NIV)

5.1 The effect of geometric spreading

When a vertical High Frequency (HF) whip aerial radio transmitter radiates wave energy equally in all directions, some of which may become a sky wave as discussed in the previous chapter, the output power will be spread equally over the surface of a three-dimensional sphere of increasing radius. The strength of the signal recorded by a radio receiver in any given direction is going to decrease as the receiver moves further away from the transmitter. A useful quantity that helps to compare signal strength, taking into account the reduction due to moving further away from the transmitter, is the wave *intensity*.

The intensity at a point can be defined as the rate of flow of wave energy per unit area, orthogonal (perpendicular) to the flow direction at any point. The unit of intensity, involving energy per second, is measured in watts per square metre, or W m^{-2}.

Figure 5.1 illustrates what is called *geometric spreading*. At the radio receiver a distance R from its corresponding radio transmitter, the wave energy will spread out uniformly over the surface of a sphere of radius R.

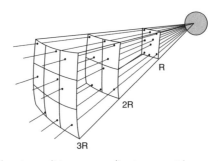

Figure 5.1: *Power spreads out over distance according to geometric spreading.*

The rate of flow of energy leaving the radio transmitter is simply the transmitted *power (P)*, usually quoted in terms of the average power level of the transmitter over a given period of time. The *Intensity (I)* at the receiver at A is thus given by:

$$\text{Intensity (I)} = \frac{\text{Transmitted power (P)}}{\text{Surface area of a sphere of radius R } (4\,\pi\,R^2)}$$

If you examine this relationship closely, you will observe that this relation shows that the intensity is:

(a) Directly proportional to the transmitter power level, P, and

(b) Inversely proportional to the square of the range, R. (i.e. the *inverse square law*). Many forces of nature, such as gravity and electromagnetism, decrease with increasing distance in this way due to the geometric spreading – 'stretching out' – encountered by the respective lines of force. So if the distance is doubled, the power will fall by a factor of 2 squared, i.e. 4.

Hence: $I = P / (4\pi R)^2$ (**eq 5.1**)

Example 5.1: Consider a man-portable backpack radio transmitter with a 100 W output power level. What is the exact intensity of the radio transmitter at a distance of 200 m?

Using the formula given above: $I = 100 / (4\pi(200)^2 = 1/\pi$ W m^{-2} as an exact solution.

Example 5.2: If the intensity 5 m from a transmitter is 10 W, what will be the intensity 17 m away (3 decimal places)?

We now need to consider two sets of conditions: $I_1 = \dfrac{P_1}{4\pi R_1^2}$ (**eq 5.2**) and $I_2 = \dfrac{P_2}{4\pi R_2^2}$ (**eq 5.3**)

Dividing equation 5.2 by equation 5.3 gives: $\dfrac{I_2}{I_1} = \dfrac{P_2}{P_1} \times \dfrac{R_1^2}{R_2^2}$ (**eq 5.4**)

If the source has not changed between the two measurements: $P_2 = P_1$ so: $\dfrac{I_2}{I_1} = \dfrac{R_1^2}{R_2^2}$ and $I_2 = \dfrac{I_1 \times R_1^2}{R_2^2}$

substituting for the values given: $I_2 = \dfrac{10 \times 5^2}{17^2} = 0.865$ Wm^{-2}

Although P_1 and P_2 can generally be considered as equal for radioactive sources, this should not be generally assumed for a radio transmitter, which can both decrease and increase its output power levels during transmission.

For an echo, waves undergo the inverse fourth power law, spreading on both the outward and return journeys. Intensity varies inversely as the fourth power of the range; consequently, intensity of echoes will be proportional to $1/R^4$.

In reality, the measured intensity for all sources is less than that predicted from geometric spreading alone. Further mechanisms of electromagnetic wave intensity reduction are present in both the above water and the underwater environment. To understand these processes for electromagnetic waves, let us consider underwater light attenuation.

In section 5.5 we will return to considering the inverse square law, which can be used to show the dominant effect of the moon on tides rather than the sun, since the change in force across the Earth will be much greater for the moon gravitationally than for the more distant sun, even though it has a much greater mass.

5.2 Underwater light attenuation

The deeper you go down or dive into marine waters, the darker they will become. Light levels weaken or attenuate with depth so rapidly that in clear waters blue-green light levels are reduced to typically just 1 per cent of its near surface value in about 100 metres, decreased to perhaps only 30 m in shallow seas. Red light level falls the most dramatically, dropping to the 1 per cent level in the first 10 m. In fact, intensity generally falls rapidly with depth right across the visible and NIR spectrum. This process of reduced signal strength is called *attenuation*.

Attenuation is actually composed of two very different mechanisms: *absorption* and *scattering*. Basically, absorption is a mechanism whereby energy is lost from the incident electromagnetic wave and is consequently converted or turned into other forms of energy – for example, the heating effect observed when food is placed within a microwave oven or a foolish sailor stands in front of a 'live' radar transmitter or puts a hand on a transmitting radio antenna. In scattering, on the other hand, variations in the propagating media will cause electromagnetic waves to alter their propagation direction from their otherwise 'straight through' path – for example, the effect observed while driving on a foggy night.

In 'pure' water you might see a visibility of 100 m range, but this is unlikely in the inshore waters off the south-west English coast, where visual range of 20 m would be considered good. If, after a winter storm, rivers such as the Dart in the south-west of England bring down suspended matter into the sea from the wilderness of Dartmoor and surrounding agricultural land, optical ranges of as little as 1 m are quite common. In muddy rivers, visible range can be as little as 0.1 m, so a diver will have difficulty seeing a hand in front of his face even if just submerged. Deeper in the sea, less light penetrates until finally human vision is impossible, a depth that can vary between 1 m in 'dirty' water up to 100s of metres when very clear.

Conditions for vision underwater on a good day are generally comparable with foggy conditions observed at sea in the above water environment. It isn't specifically the low light intensity that prevents distant object viewing above or below the waterline, but rather the light scattering from water droplets in fog, which degrades contrast below the minimum required for identifying an object against its background. This is not surprising as a medium can be translucent (light can pass through it) but not transparent (clear enough to see an image).

Absorption is quite selective and varies in both its amount and its spectral distribution across the world's oceans, depending on their apparent 'water quality'. Water molecules have generally broad absorption resonances in the visible and infrared bands. However, a process of selective absorption also occurs.

Of all the visible and NIR wavelengths, it is the blue-green wavelengths that propagate the best in seawater, with the lowest loss. Under appropriate environmental conditions, strongly dependent upon scattering and absorption, underwater imaging, exploration, and even the surveying of wrecks are possible. Red underwater lasers are used increasingly in research applications but, for detection of underwater mines, blue-green lasers have superior performance – but also have the drawback that they are more expensive.

5.3 Absorption of underwater light

When a photon of light interacts with a water molecule it may be absorbed, with the energy being transferred to the water molecule. Light absorption is quantised and energy can only be absorbed in discrete sized steps or *quanta* because energy content of an atom or molecule is limited to just a discrete numbers of levels, such that:

$$E_n - E_{n-1} = hf \qquad \textbf{(eq 5.5)}$$

where E_n and E_{n-1} are next nearest related energy levels.

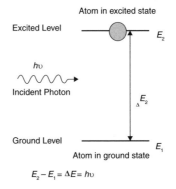

Figure 5.2: *Quantisation of energy.*

Example 5.3: For a molecule raised from its normal state of 2.5×10^{-15} J to 5×10^{-15} J, what is the frequency of radiation emitted when a photon is emitted as the molecule returns to its ground state (3 significant figures)?

Using equation 5.4, $E_n - E_{n-1} = hf$

$$5 \times 10^{-15} - 2.5 \times 10^{-15} = hf$$

So: $f = \dfrac{2.5 \times 10^{-15}}{6.6 \times 10^{-34}} = 3.79 \times 10^{18} Hz$

Atomic absorption: Electrons also have precise energy levels so that only certain specific *orbits* are possible. Atoms absorb energy in discrete or fixed amounts, comparable with the discrete increases in potential energy that result from walking up a staircase one step at a time. If sufficient energy is available to raise an electron beyond the next available orbital, the electron can move to the next orbital and a quantum of energy is absorbed. A molecule can actually have many electronic energy levels associated with its component atoms and also has other available quantised vibrational and rotational energy levels – such is the wondrous complexity of our universe at the molecular, atomic and subatomic levels.

However, absorption of visible light by water generally involves low energy molecular vibration rather than high energy electronic level changes. Nonetheless, these low energy absorptions occur in the visible spectrum, which is still high in frequency compared with radar frequencies. Water selectively absorbs low energy (long wavelength) red, and consequently the low absorption of higher energy, low wavelength (short wavelength) blue light helps generate water's characteristic blue colour. The absorption coefficient A (λ) values for pure water in the visible spectrum are given in Table 5.1. B (λ) is the scattering coefficient and C (λ) is the total loss coefficient. Some comments on the properties of seawater are given in Appendix B.

λ nm	Colour	A m⁻¹	B m⁻¹	C =(A+B) m⁻¹
410	Violet	0.016	0.007	0.023
470	Blue	0.016	0.004	0.020
535	Green	0.053	0.002	0.055
555	Yellow-Green	0.06	0.002	0.069
575	Yellow	0.094	0.002	0.096
600	Orange	0.244	0.001	0.245

Table 5.1: *Absorption coefficient A(λ), B (λ), and C(λ) values for pure water at visible wavelengths.*

In water, absorption is found to be minimum at the short wavelength blue end. The complex distribution of dissolved salts that are found in seawater actually have relatively little effect upon its overall optical properties so the values given here can be taken as relatively accurate for seawater, but of course will have different and significant corrosive properties. This is potentially a critical factor when considering the different dissolved salt compositions in the various oceans of the world, which are not identical for different immersed metal alloys.

Using instruments more sensitive than the human eye, satellites measure ocean colour and can be used to track sea surface temperatures and currents. Mapping ocean colour reveals the presence and concentration of marine animals, phytoplankton, sediments and dissolved organic chemicals. A Sea-viewing Wide Field-of-View Sensor (SeaWiFS) provides qualitative data on global ocean bio-optical properties (figure 5.3 – see plate section); such images can tell us a lot about the state of the ocean. Mariners use remote sensing satellites to measure various parameters to provide total ocean surveillance on a global scale. Each data point gathered represents an individual unit or parcel of Earth's surface. New satellites and new sensors mean there is an ever increasing amount of data collected.

5.4 Scattering of underwater light

When a photon (packet) of electromagnetic energy interacts with a molecule, it may be scattered many times prior to absorption. This process will result in the photon changing its overall propagation direction. Scattering doesn't reduce the photon's energy but will, however, increase its probability of encountering an absorbing molecule as its path through the absorbing medium is extended by comparison with its direct unscattered counterpart.

There are two important common types of scattering:

1) Rayleigh scattering

2) Mie scattering

5.4.1 Rayleigh or molecular scattering

If an air molecule is placed within a beam of electromagnetic wave energy such as light or ultraviolet radiation, a dipole will be induced by the incident time-varying electric field vector, causing it to radiate. As the dipole oscillates at the exciting radiation frequency, it re-emits radiation at the same frequency in all directions. This re-radiation is the scattered light, and this may be considered as a microscopic antenna of molecular dimensions.

Scattering is found to be proportional to λ^{-4}, where λ is the wavelength of the light, and is thus greater at short wavelengths.

Scattering is also found to be proportional to d^{-2} where d is the molecule diameter, and is found to be greatest for the smallest molecules. Scattering is uniform in all directions, i.e. the wavelength is scattered isotropically. In fact, there is a pioneering scientific technique used to determine the chemical composition of materials in containers and behind a range of other barriers, including skin, which may lead to future applications including real-time diagnostic tools for cancer and bone disease. The so-called RAMAN effect is where a beam of light hits a material and a small number of the incident photons interact with the material's molecules, either gaining or losing energy. Every material produces slightly different energy changes, so by measuring this scattering 'fingerprint' it is possible to identify the substance being tested.

Example 5.4: What is the integer ratio of scattered light at 400 nm to that scattered at 800 nm?

$$I_s \propto \frac{1}{\lambda^4}$$

Then: $I_{400 \text{ nm}} \propto \dfrac{1}{400^4}$ **(eq 5.6)** $I_{800 \text{ nm}} \propto \dfrac{1}{800^4}$ **(eq 5.7)**

Dividing equation 5.6 by equation 5.7: $\dfrac{I_{400 \text{ nm}}}{I_{800 \text{ nm}}} = \dfrac{800^4}{400^4} = 2^4 = 16$

Example 5.5: If the scattering at 400 nm is 0.5 Wm^{-2}, what is the intensity of scattered light at 800 nm (5 decimal places)?

Since $\dfrac{I_{400\ nm}}{I_{800\ nm}} = 16$ $I_{800\ nm} = \dfrac{I_{400\ nm}}{16}$ so $I_{800\ nm} = \dfrac{0.5}{16} = 0.03125$ Wm^{-2}

The scattering coefficient B(λ) m^{-1} of seawater is given for wavelengths in the visible range in Table 5.1, as expected for λ^{-4} Rayleigh-type scattering. B(λ) is a minimum in the red and a maximum in the blue part of the spectrum. It is the high scattering of blue combined with low absorption of blue A(λ) that combined to yield C(λ), which gives clear seawater its usual blue colour.

Rayleigh scatter is also responsible for the blue colour of the daytime sky, as shorter wavelength blue wavelengths will be scattered preferentially out of the direct path between the top of the atmosphere and the Earth's surface. This process is most exaggerated at sunrise and sunset, where direct unaided views of the sun by the human eye will observe a solar source that is significantly 'reddened' due to the loss of the blue, shorter wavelength component from our exaggerated optical path.

5.4.2 Mie scattering

While molecular scatterers are small relative to such visible light wavelengths, most light scattering in water is from particles much bigger than about 2 µm diameter, i.e. particles that are very large compared with the incident visible wavelengths. Marine and river water will usually have a high concentration of such sized particulates. *Diffraction*, reflection and refraction contribute to scattering from these large particles. Diffraction will be determined by the size and shape of complex scatterer structures, and we will look at this phenomenon in a little more detail in Chapter 6, while refraction and reflection depend upon the medium's refractive index, as has already been discussed. The basis for predicting light scattering behaviour of any size is quite complex and is given by Mie theory. Unlike symmetrical Rayleigh scattering, particles of large size predominantly scatter in the propagation direction, i.e. in the *forward direction* rather than back towards the light source.

We have already discussed geometric spreading, although it should be emphasised that in some real practical applications, such as sonar and radar ducting, the detected signal strength equations need to be modified further. This is because the inverse square law does not apply due to the trapping of radiation, which is then able to travel a greater distance than normal as it is guided down a duct in a similar way to light being guided down an optical fibre. It should also be noted that a radar echo will not only suffer absorption and scattering on the way out to the radar target but will also suffer further geometric spreading losses on the way back, such that the intensity of any detected echoes will be inversely proportional to the 4th power of range (R), i.e. $I = k/R^4$ where k is constant.

This same principle of ducting or radiation guiding is also responsible for the great distance of optical fibre propagation, where light can be bound or constrained to travel for up to 50 km or more before reaching the next repeater station where the signal to be sent is retransmitted, without errors being introduced. However, this requires optical glasses that are extremely pure, unlike the glass used in typical glassware and in windows, which are too absorbing to be used for such applications. (Note: Take a piece of ordinary glass and turn it on its side – you will see that the glass is usually green, due to the presence of copper and iron impurities, figure 5.4 and 5.5 – see plate section for both).

Whenever there is a medium present, the effects of absorption and scattering must be included and are often considered together under the generic term of *attenuation*. In the case of underwater optics discussed above, the absorption coefficient $A(\lambda)$, and a scattering coefficient $B(\lambda)$ simply add together to give the overall attenuation coefficient in a linear additive manner:

$C(\lambda) = A(\lambda) + B(\lambda)$ **(eq 5.8)**

The direct consequence of the presence of absorption and scattering is that the reduced measured intensity will fall to a level that is less than that predicted simply from considering the inverse square law equation discussed earlier.

E.g., for water at a 470 nm wavelength where $A = 0.016$ m^{-1},

and $B = 0.004$ m^{-1} then $C = 0.016 + 0.004 = 0.020$ m^{-1}

Example 5.6: If water depth is 10 m and surface incident light intensity in a collimated beam $I_{in} = 1$ Wm^{-2}, what is the light intensity I_{out} at 10 m depth (3 decimal places)?

Using: $I_{out} = I_{in} e^{-C(\lambda)x}$ **(eq 5.9)**

$I_{out} = 1 \times e^{-0.020 \times 10} = 0.819$ Wm^{-2}

5.5 Additional comments on the inverse square law regarding tides

Let us consider finally the occurrence of tides in context of the inverse square law. It is known that tides are affected by both the sun and the moon. One could ask why does the moon have a greater effect on the tides than the sun? Although the sun is about 390 times further away from the Earth than the moon, its force on the Earth is about 175 times greater, yet its tidal effect is smaller than that of the moon's. This is because tides are a result of differences in gravitational attraction across the planet. As Earth's diameter is so small in relation to the sun's, there is hardly any variation in its gravitational attraction across Earth's surface, while the moon's diameter is significantly less, therefore there is a much greater variation in gravitational attraction across Earth's surface – which is an example of the inverse square law.

In the case of gravity, if the sphere has an area of $4\pi R^2$ and the source strength is $4\pi GM$, so the intensity at the surface of a sphere will be given by using equation 5.1 earlier: $I = P / (4\pi R)^2$

$I = P / (4\pi R)^2 = 4\pi GM / (4\pi r)^2 = GM / r^2 = g$, the acceleration due to gravity arising from the moon.

When the radius of the Earth is doubled, this value falls to $g/4$.

Consider increasing the distance from the Earth by a small distance dr then the overall change in gravitational force dF can be found from the Force equation: $F = 4\pi GM / (4\pi r)^2$ then $\dfrac{dF}{dr} = -2GM / r^3 = \dfrac{-2g}{r}$

So $dF = = \dfrac{-2g}{r} dr$

Substituting for known values of dr (approximate Earth diameter) = 12742 km and with r = 360 000 km, so the gravitational attraction of the moon at its surface is: $dF = -0.06888$ F_{MOON} where F_{MOON} is the gravitational attraction of the moon g at the surface.

Similarly, the difference in gravitational attraction of the sun across the Earth is:

$dF_{SUN} = 0.00017 \times F_{SUN}$ where dF_{SUN} is the change in gravitational attraction across the planet and F_{SUN} is the gravitational attraction of the sun at its surface – thus demonstrating that the moon's tidal effect is greater than the sun's tidal effect.

It is also possible to prove this using differentiation from first principles but this takes much longer and is much more tedious. It is left to the reader to prove this.

5.6 Self-assessment questions

After studying this chapter, you should be able to answer the following questions.

1.   State the relationship between transmitted power and the intensity at any range from the transmitter.

2.   A radar transmitter has a 25 W power source. If the source radiates isotropically in all directions, what will the intensity be 60 m from the source (3 significant figures)?

3.   A radar set detects a target at a range of 75 km. The intensity of the signal striking the target is 14×10^{-2} W m^{-2}. Calculate the new intensity if the range is decreased to 55 km and the transmitted power is increased by 30 per cent (3 decimal places)?

4.   For echoes, waves spread out both on transmission and on the return echo path, so what will the echo intensity as a function of distance now look like?

5.   For a target at a range of 50 km and the same target at a range of 150 km, what will be the ratio of the detected echo strength (5 decimal places)?

6.   A radar set can just detect a target at a maximum range of 110 km. If the power output of the transmitter is increased by 25 per cent, what will be the new maximum range of the radar in km (4 significant figures)?

7.   What is the ratio of scattering intensity for a wavelength of 450 nm compared with a wavelength of 650 nm (3 decimal places)?

8.   Using the beam attenuation coefficient C(λ) at 575 nm and for a water depth of 20 m and incident light intensity I_{in} = 25 Wm^{-2}, find the outgoing light intensity I_{out} (2 decimal places).

9.   Intensity is proportional to both d^{-2} and λ^{-4}. When dI/dd = dI/dλ, what will the value of wavelength be in terms of scatterer size?

10.   For water at a 535 nm wavelength, with a water depth of 15 m and surface incident light intensity I_{in} = 0.75 Wm^{-2}, what is the light intensity I_{out} at 13 m depth (2 decimal places)?

6

Diffraction

'Everything we see is a shadow cast by that which we do not see.' Martin Luther King Jr

6.1 The effect of an aperture

When radiation passes out of a radar transmitter or past an obstruction, waves can be engineered to change direction. This bending of light is different to refraction, which relies upon a difference of refractive index between two media. Bending of radiation in this case, whether from a radar or radio source in space, occurs at the *edges* of an opening or aperture, or from an obstruction. The result of this is that radiation will reach regions that would otherwise remain in the object's geometric shadow. For example, when fierce January Atlantic storm waves are breaking in Plymouth Sound on the English Channel, a surfaced submarine will experience considerably turbulent rocking motions while in harbour, even if it is well within the *geometric shadow* of the harbour wall. In figure 6.1, waves of wavelength λ are observed to pass through an aperture of width D. Experiment shows that the amount of bending is found to be inversely proportional to the aperture width.

If the aperture width is changed while maintaining the same wavelength, a smaller value of *beam width* (α) and a narrower beam spreading will be produced after passing through the opening. In the limit of an infinitely narrow source, or point source, the source will radiate freely in all directions uniformly (or isotropically). Increasing aperture width infinitely, on the other hand, would produce parallel waves that exhibit no bending (figure 6.1). Thus it is the edge effects that are responsible for the spreading or diffraction of incident radiation.

The degree of bending is found to be directly proportional to the wavelength (or inversely proportional to frequency). It is noted that although we have clearly considered a horizontal slit or aperture, the same diffraction phenomenon is responsible for vertical spreading of a beam as well. In this way, a radar engineer can independently design horizontal and vertical spreading beam widths for a variety of applications to create radar with very different shapes, such as those for navigation or air and surface search. This spreading by a single slit is thought of by considering *Huygen's theory*, which treats every point along the wave front passing through the aperture as separate secondary sources of waves, giving rise to interference effects, which we will not discuss in depth but can be explored elsewhere [6.1].

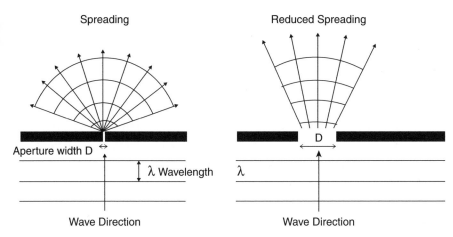

Figure 6.1: *Diffraction changes as aperture size changes.*

6.2 Optical diffraction

Diffraction has been investigated extensively at optical and microwave frequencies. Figure 6.2 shows the effect of a beam of *monochromatic* (single wavelength) light from a laser passing through a very narrow slit. When the intensity of the visible light is measured along a line on a screen positioned some distance behind the slit, dark and light bands are observed to have an intensity that varies with the diffraction angle θ, as shown.

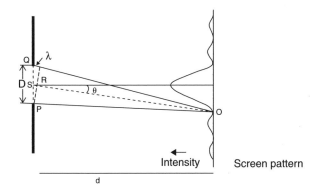

Figure 6.2: *Laser optical diffraction pattern through a slit on to a screen.*

Consider the waves coming together at the first minimum to the left or the right of the central path's direct straight-through beam. For waves to come together at these points, waves from the centre and half an aperture either side must be half a wavelength out of phase (destructive interference is needed).

The intensity is a maximum, I_{max}, in the 'straight-through' position and decreases, as might be anticipated, to zero as the detector moves further away from this central position, i.e. as the angle θ increases. However, on increasing the angle θ still further, it is observed that the intensity is found unexpectedly to rise again to a value less than I_{max} before falling once more to zero, i.e. further 'secondary' maxima occur well into the expected geometric shadow of the slit. The effect is known as *single slit diffraction* and the shape of the intensity profile can be predicted theoretically. Similar effects occur with multiple slit sources. Single slit (figure 6.3, see plate section) and multiple slit (figure 6.4, see plate section) can be compared side by side with incident Helium-Neon (He-Ne) 632.8 nm laser radiation.

6.3 Diffraction at a slit

Considering figure 6.2, if the angle θ is such that the path difference OS-OP is $\dfrac{\lambda}{2}$

then OS-OP waves will interfere destructively. Consequently, no light will be seen in the direction θ.

For a dark fringe $\dfrac{d}{2}\sin\theta = \dfrac{\lambda}{2}$

Thus in triangle SPR $\sin\theta = \dfrac{\dfrac{\lambda}{2}}{\dfrac{d}{2}}$

i.e. $\sin\theta = \dfrac{\lambda}{d}$ (**eq 6.1**)

or since θ is very small, $\sin\theta$ closely approximates to θ in radians.

6.4 Diffraction gratings

Diffraction gratings consist of many narrow, regularly spaced openings, and are produced in flat metallic coatings or with repetitive saw wave or sinusoidal surface profiles, essential for a wide variety of modern electro-optic components and functions. However, grating studies exist widely in nature as well and much research is currently underway to replicate the complicated structures of butterfly wings [6.2] and beetles [6.3] for optical applications as well as materials with structural light interference, such as cellulose. Butterflies such as the iridescent blue morpho butterfly (figure 6.5, see plate section) contain non-metallic gratings that give very strong reflections of blue light at precise angles of incidence.

Such gratings may in future provide interesting sensor elements as their behaviour changes dramatically if the environment varies near the grating. For example, if pure air (n = 1.0003) on the butterfly's wing in the bottom right section is replaced with a thin coating of acetone (refractive index = 1.3590), the wavelength best reflected (brightest observed reflection) changes in a dramatic and startling manner from blue to green (figure 6.6, see plate section) [6.4].

Now, if parallel light falls on the grating, the result is that the wave front is split into many separate coherent sources. In some directions, the path differences between adjacent sources is λ and so the waves from all the sources across the grating will be *in phase*. In these directions, a bright image is formed. The condition for two adjacent sources to be in phase and for a bright spot on a screen is that d $\sin\theta$ = n λ, where n is an integral number of wavelengths and d is the distance between regularly ruled openings in the diffraction grating; n also gives the *order* of the image (figure 6.7). White light splits into a number of spectra, since each colour exits the grating at a different angle (see figure 6.8, see plate section). The red wavelengths are diffracted and arrive at the screen in phase with a small value of θ displaced a short distance above (or below) the straight-through beam. The visible shorter violet wavelengths are displaced further in angle and vertical distance before again coming together in phase.

Note: The condition for a diffraction grating for constructive interference appears the same as that for a single slit, resulting in destructive interference. Remember the d here is the spacing between individual openings in the grating and does not refer to the slit width!

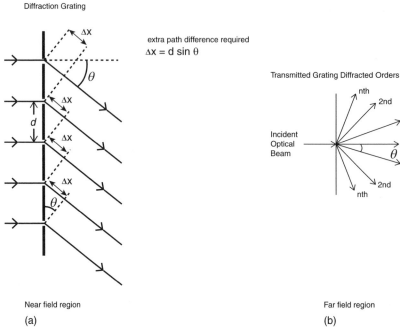

Diffraction Grating

extra path difference required
Δx = d sin θ

Transmitted Grating Diffracted Orders

Near field region

(a)

Far field region

(b)

Figure 6.7: *Diffraction grating in-phase conditions and orders.*

Example 6.1: A diffraction grating of 5000 lines per centimetre is used with white light. How many orders of spectra will be observed? Consider figure 6.8.

No. of lines per metre $= 5 \times 10^5$

Spacing $d = \dfrac{1}{5 \times 10^5} = 2 \times 10^{-6}$

For the *nth* constructive order for a grating:

$d\sin\theta = n\lambda$

The maximum value of θ incident is 90°, so sin (90°) = 1

Taking the spectral range for white light to be between 400 nm and 700 nm, the mid-point of the visible spectrum observed will be about 550 nm or 550×10^{-9} m

So the maximum possible value of n is:

$n = \dfrac{d\sin\theta}{\lambda} = \dfrac{2 \times 10^{-6} \times \sin 90}{550 \times 10^{-9}} = 3.64$ (2 decimal places)

Therefore, three complete orders should be seen.

6.5 Diffracted beam shape at a radar aperture

Taking the intensity profile seen already in figure 6.2, intensity can be recorded as a function of angle of the normal straight-through direction (figure 6.9).

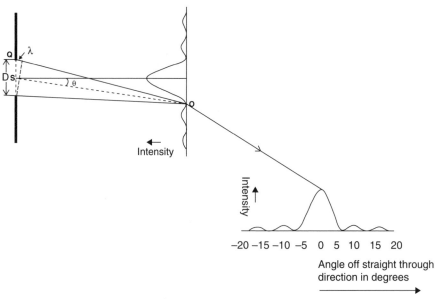

Figure 6.9: Angular diffraction.

The diffraction angle at which the first intensity zero occurs depends upon the precise shape of the diffracting slit or obstacle, but for practical purposes it can be taken to be roughly:

Beam width $a = 60\lambda/D$ **(eq 6.2)**

where D is the aperture opening (width or height respectively for the horizontal and vertical beam width respectively). The factor of 60 comes from an approximate radians to degrees conversion. Some radar are best approximated by a factor of 55 or in some cases as high as 70, but 60 is a simple factor to use, and generally close enough to any real radar's output for practical general use.

Example 6.2: For a radar with a horizontal aperture of 2.1 m and a wavelength of 10 cm, find the resulting horizontal beam width to 2 decimal places using equation 6.2 (2 decimal places).

Using $a_{horizontal} = 60\lambda/D_{horizontal}$ **(eq 6.3)**

And substituting so: $a_{horizontal} = 60 \times 0.1 / 2.1 = 2.86^0$

This is nearly identical to the angular width of the main beam if it is measured between the half intensity or half-power points of the curve (figure 6.10). This angular range or width is thus commonly referred to as the *beam width*.

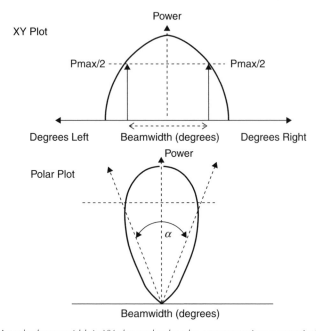

Figure 6.10: *Angular beam width in XY plot and polar plot representations respectively.*

In radar and radio antennae, just as with visible light diffraction, diffraction occurs at the antenna's edges and the consequent radar main beam widths in both the vertical and horizontal directions respectively are controlled by both the overall dimensions of the antenna and the frequency (wavelength) at which the radar operates,

i.e. $\alpha_{vertical} = 60\lambda/D_{vertical}$ (**eq 6.4**) and $\alpha_{horizontal} = 60\lambda/D_{horizontal}$

(given as equation 6.3 previously).

The secondary maxima that also occur are a problem for good radar operation and are known as *side lobes*. Large side lobes are a waste of radar power that should instead be in the main beam of the radar. Carelessly designed characteristic radar side lobe patterns also allow the identification of specific radar systems. Military radar side lobes are also vulnerable to the action of radar jamming. Modern phased array radar are designed to have very small side lobes due to variations in the output power levels of individual radar receivers (or sources) to mitigate against such threats.

6.6 Arrays of multiple sources

The full subject of multiple source interference relevant to advanced radar systems will be considered in a little more detail in Chapter 7 under the topic of *interference*. However, here we will treat the output of multiple regularly spaced radar sources by first considering two regularly spaced sources. This is the same as the principle of interferometry used in radio astronomy.

Consider two radio telescope dishes, A and B, separated by a distance of perhaps 3 km (figure 6.11, see plate section). Radar energy from a distant celestial source will reach these two radio (or radar) dishes. As encountered previously, if the received path difference is a whole number of wavelengths the two signals will be in phase and a maximum output signal will result (*constructive interference*), but if the two signals are out of phase the resultant signal will be a minimum (*destructive interference*). So as an astronomical source moves across the night's sky, the resultant signal will be observed to vary in intensity between these two extreme limits. This is a similar problem to Young's two-slit experiment, except that we are now considering two point receivers instead of two emitters of radiation. The mathematics of *reciprocity* will, however, mean the outcome will be the same in both cases.

The variation in intensity with angle will appear similar to the variation in interference fringes seen in Young's two-slit experiment. Successive maxima or peaks are thus obtained at an angular separation equal to λ/d, where λ is the wavelength and d is the separation of the two dishes. So if the wavelength is 5 cm and $d = 3$ km then the angular separation $= 60\lambda / d = 60 \times \dfrac{0.05}{3 \times 10^3} = 0.001\,\text{degrees}$.

The angular resolution θ, the angle from the central maximum to the first minimum, is given by $\theta = \dfrac{\lambda}{2}$.

The Very Large Array (VLA) in New Mexico (figure 6.12, see plate section) consists of 27 radio telescopes each 25 m in diameter, arranged in a Y-shaped configuration. All 27 telescopes can thus be used simultaneously to observe a target and then their individual observations can be added together. The greater the distance between any two telescopes, the better the resolution we will achieve when they are used together. Radio astronomers have used telescopes thousands of kilometres apart on different locations on the Earth's surface and they have even been placed far apart in space to improve resolution further. This method is called *Very Long Baseline Interferometry* (VLBI). Each telescope has its own atomic clock and records its separate observations. Later, observations from the different telescopes can be synchronised and then combined. For example, consider a single 25 m diameter dish detecting 5 cm waves, the narrowest beam width achievable $= 60\lambda/D = 60 \times 0.05 / 25 = 0.12°$ with $D = 25$ m.

However, consider replacing a single dish with nine identical dishes separated by 100 m each. The total length of the array, D, is now 800 m (eight spaces between nine receivers). This results in a new beam width of $60\lambda/D = 60 \times 0.05/ 800 = 3.75 \times 10^{-3}$ degrees, a significantly better value than that achieved by having one dish on its own. In the same way, the use of a regular array of radar sources will produce a highly focused and directional beam of energy. Indeed, an array can be used to enhance performance in this way at any electromagnetic wavelength, and can also be used to enhance sonar performance of an acoustic array.

6.7 Self-assessment questions

After studying this chapter, you should be able to answer the following questions.

1.   Sketch diagrams to show the effect of different aperture widths on diffracted radiation. Explain the reasons for the first secondary maxima in the geometric shadow of a single slit.

2.   The horizontal and vertical beam widths of a radar antenna are 20° and 1° respectively. The radar operates at 9.5 GHz. Calculate the horizontal and vertical dimensions of the antenna (3 decimal places).

3.   The vertical and horizontal dimensions of a radar antenna are 25 cm and 2.4 m respectively. If the radar operates at 3.1 GHz, what are the horizontal and vertical beam widths (2 decimal places)?

4.   Explain the process of single slit diffraction giving rise to the first detected minima in transmitted angle.

5.   Explain the process of multiple slit diffraction.

6.   A diffraction grating of 6000 lines per centimetre is used with white light. How many orders of spectra will be observed?

7.   Explain how an array of multiple receivers can improve angular detected resolution.

8.   With two receivers, calculate the resolution limit if the wavelength is 2 m and the two receivers are spaced 1 km apart.

9.   Calculate the resolution limit if the wavelength is 2 m and there are now 40 receivers spaced across the 1 km distance of question 8.

10.   What is the receiver spacing in metres if a linear array can resolve a beam width of 0.001° if the wavelength is 5 cm with 60 elements in a sonar array (2 decimal places)?

References

[6.1] *Treatise on Light*, C Huygens (1690; translation published Dover Publications, New York, 1962).

[6.2] 'Quantified interference and diffraction in single Morpho butterfly scales', P Vukusic, JR Sambles, C Lawrence and RJ Wooton, Proc. R. Soc. Lond. B, 266 (1999), pp. 1402–1411.

[6.3] 'Photonic structures in biology', P Vukusic and JR Sambles, *Nature* Vol 424 (August 2003), pp. 852–855.

[6.4] 'Thermal analysis of butterfly wings', CR Lavers and S Shepard, ed. N Lamontagne, Biophotonics International (February 2008), pp.84–85.

7

Interference

'A day on the beach with the family is usually better than the best day in the office.' Christopher Lavers

7.1 Principle of superposition

It is observed that when water or electromagnetic waves from more than one source arrive at a point (such as on a beach), the resultant action at that point in space is equal to the sum of effects that would have occurred at that point if each of the sources had been acting alone. This is described as the principle of *superposition*, and means that the overall effect at the chosen point depends on the individual amplitudes, frequencies and relative phases of all the arriving waves.

This somewhat surprising result provides opportunity for many useful wave devices and effects. One particular case is that of modern active phased array radar, whose large overall power output is achieved by the use of hundreds or several thousand relatively small power transmitters. An additional benefit of this design architecture is that if one of the individual power module emitters fails, the other sources can easily compensate for this loss. This type of system degradation failure is usually called *graceful degradation*. A radar system with only one power source that stops working will suffer a catastrophic failure, and all its functionality will be lost.

Often, the transmitted and received waveforms that occur in space and over time can be quite complicated. However, it can be shown from the use of Fourier analysis that any complex waveform can be analysed and broken down into a sum of simple waves of *harmonically* related frequencies, having different amplitudes and phases. This is an extremely important result for mathematics, physics and engineering.

The subject really began with the study of the way general mathematical functions can be represented by sums of simpler trigonometric functions. Fourier analysis is named after Joseph Fourier (1768–1830), a French mathematician who showed that representing a function by a trigonometric series greatly simplifies the study of heat propagation in metal rods. Today, Fourier analysis encompasses a wide spectrum of mathematics. In science and engineering, the process of *decomposing* a function into smaller, simpler parts is called *Fourier analysis*, while the operation of constructing the function from these smaller building blocks is known as *Fourier synthesis*. In mathematics, the term *Fourier analysis* often refers to the study of both these processes.

We can draw diagrams to represent complex waves. Sometimes the *time domain* representation (waveform) is used (figure 7.1) but sometimes the *frequency domain* (frequency spectrum) is useful instead (figure 7.2), remembering that the frequency spectrum is the inverse of the time domain ($f = 1/T$).

Displacement/m

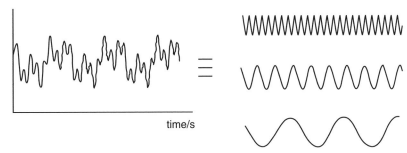

time/s

Figure 7.1: *Time domain.*

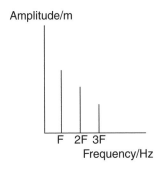

Figure 7.2: *Frequency domain.*

In figure 7.1 we see the time domain of a waveform that exists as the superposition of three harmonically related sine waves. In figure 7.2, in the frequency domain, we can see the three harmonically related waves and their relative amplitudes. By the use of appropriate mathematical techniques, the engineer can transform from one representation to the other.

An example of Fourier analysis: Consider wanting to construct a square wave with an equilibrium about x = 0. We need to add together a series of harmonically related frequencies of different amplitude, which for a square wave is composed of all the odd harmonically related integer frequencies.

$$\text{Square wave} = \sum_{\substack{n=1 \\ n\ odd}}^{N} \frac{1}{n}\sin n\ \omega_0 t \qquad \textbf{(eq 7.1)}$$

So let us consider a simple example with a fundamental frequency (n = 1) with ω_0 = 50 Hz (so-called mains frequency). Then the first harmonic frequency will be 100 Hz and the second harmonic frequency 150 Hz and so on. Considering the first three terms only:

Signal $(\omega) = 1 * \sin(\omega_0 t) + (1/3) * \sin(3\omega_0 t) + (1/5) * \sin(5\omega_0 t)$

These frequencies and their summative addition are seen in the corresponding Excel Graph (figure 7.3, see plate section)

In this figure, the reader can observe the fundamental frequency displayed by series 1. The first harmonic frequency, existing at twice the fundamental frequency, is given by series 2. The combination of series 1 and series 2 gives rise to series 4, which now begins to look less like a sine wave. The addition of the second harmonic (at three times the fundamental frequency) is observed in series 3, which combined with series 1 and series 2 creates the overall Fourier result of series 5, which is now a passable square wave, constructed solely from harmonically related frequencies.

This also demonstrates that the bandwidth of a square wave is considerably greater than the fundamental frequency alone, which has the same period as a square wave but certainly does not have its shape (constructed as it is from the addition of the harmonically related frequencies).

7.2 Same frequency interference

When two (or more) waves having the same frequency meet at a point, the waves combine according to their relative phases and amplitudes (principle of superposition). If they are in phase, they add and the resultant has a maximum displacement. Maximum intensity occurs at this point (*constructive interference*) and is known as an *in-phase condition* and will be found in radar beams or *lobes*.

If, however, the waves are in anti-phase (180 degrees out of phase), where the maximum of one wave coincides with the minimum of the other wave, they subtract and minimum intensity results. In radar terminology this is a *null*, and almost complete cancellation of waves will occur. When sources radiate waves of equal amplitude, reception in the anti-phase condition zero intensity (*destructive interference*) will result instead. Figure 7.4 shows the effect for waves of both equal and unequal amplitudes, and in-phase and out-of-phase conditions for waves of the same frequency. *Rogue waves* are often caused by the momentary formation of constructive interference of many small waves.

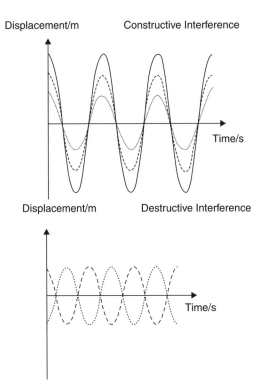

Figure 7.4: *Constructive and destructive interference.*

It follows that in the space surrounding the two sources there will be some regions where the intensity is a maximum and others where the intensity is a minimum, and in between these points there will be a huge number of points of intermediate combination values, which results in an interference pattern forming.

Results are shown in figure 7.5 (see plate section) for two sources spaced many wavelengths apart.

This pattern is completely unsuitable for a radar as there are far too many lines of constructive interference, resulting in many simultaneous radar beams. Half of the radar beams are also in the 'forward' direction, while half point in the 'backward' direction. To reduce the number of radar beams or *lobes*, the distance between sources is reduced to less than one wavelength separation and a reflector is introduced to remove the backward travelling beams, which helps to channel backward energy into the forward travelling beams.

Placement of the reflector is critical. On first thought, any multiple wavelength interval spacing would seem appropriate, but for physical antenna stability purposes it isn't practical to have many whole wavelength spacings. A large radar dish balanced on the end of a long 'boom' will have a large moment (force × distance). It would thus appear that a half-wavelength spacing would be the best minimum distance alternative, given that the round trip physical path back and then forwards for the beam is two times half a wavelength. However, on further inspection it is noted that a direct reflection introduces a half wavelength or 180-degree phase shift and so an out-of-phase condition now results.

Consequently, the desired physical distance must take this reflector phase shift into account. Look at figure 7.6 – let the reflector be placed a distance x away equally from both sources, and read example 7.1.

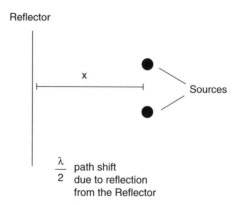

Figure 7.6: *Two sources with reflector placed behind at distance x to get an in-phase reflected wave.*

Example 7.1: What is the minimum distance a reflector should be placed behind two sources, as shown in figure 7.6, to achieve constructive interference?

Consider a wave emitted towards the reflector, which comes back in step or in phase with the transmitted wave already heading away from the source to the right.

$$2x + \frac{\lambda}{2} = \lambda \qquad \textbf{(eq 7.2)}$$

Direct out and back + mirror reflection equivalent phase shift

So: $2x = \frac{\lambda}{2}$ and thus: $x = \frac{\lambda}{4}$

Placing the reflector $\lambda/4$ behind the two emitting sources along the central line will actually provide a stronger forward travelling wave, as the original forward travelling wave is reinforced by the secondary reflected wave. However, we have not discussed the size or shape of the radar beam, which for two sources separated by half a wavelength is not only the source spacing d but also the physical aperture between the source extreme limits.

Using the beam width equation discussed earlier, i.e. beam width $a = 60\lambda/D$, if the inter-source separation d is equal to D – the actual extent of the *emitting zone* taken in a straight line between the two sources – then $D = \lambda/2$, so substituting for D in the beam width equation will result in a beam width value of 120 degrees, which is unacceptably large. Consequently, the use of an aperture of more than half a wavelength is needed, but we cannot increase the source spacing beyond the wavelength λ or we will have multiple beams and this only gets us down to 60 degrees anyway, making this unsuited for navigation and other high-resolution applications such as military fire control.

The solution (or an acceptable cheat, perhaps) is the use of a *phased array*, an arrangement of regularly spaced sources in two dimensions separated from adjacent sources by less than a wavelength, with the overall dimension of the entire array adding up to much more than a wavelength overall. We have already encountered this idea in arrays of multiple sources used in astronomical and radar applications.

Example 7.2: Consider a regularly spaced square array of sources in both the vertical and horizontal direction. If there are 10,000 sources in total and the sources in every row and column are spaced by half a wavelength, what will be the minimum resulting beam width if the wavelength is 3 cm (2 decimal places)?

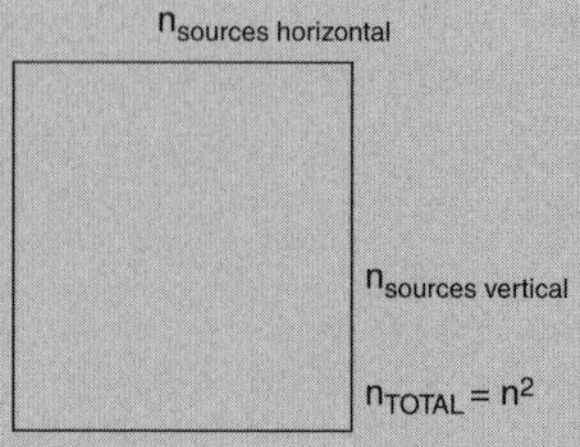

$n_{\text{sources horizontal}}$

$n_{\text{sources vertical}}$

$n_{\text{TOTAL}} = n^2$

Figure 7.7: *Square of sources with the same number vertically as horizontally.*

The total number of sources is the square of the number of sources along any row or column, i.e. the number of sources n in any row or column is related to the total number of sources N in a square array by the equation: $N = n^2$

So $n = \sqrt{N}$

Hence $n = \sqrt{10000} = 100$

Now, the number of spaces is related to the number of sources by the equation:

number of spaces = number of sources − 1 (**eq 7.3**)

So the number of spaces = 100−1 = 99

Each space d has a value of $\dfrac{\lambda}{2} = \dfrac{3}{2} = 1.5$ cm

The beam width equation gives: $a = \dfrac{60\lambda}{D} = \dfrac{60\lambda}{(n-1) \times d}$ (**eq 7.4**)

$a = \dfrac{60\lambda}{99 \times \frac{\lambda}{2}} = \dfrac{60}{99} = 1.21$ degrees, which can be found by considering the array as either an

assembly of individual sources or as a single aperture of fixed width D using the standard beam width equation: $a = 60\lambda/D$

Now, phase difference for a given path difference is given by considering an equation we encountered earlier (**eq 1.6**):

$\dfrac{t}{T} = \dfrac{d}{\lambda} = \dfrac{\phi}{360}$ (1.6 repeated)

Considering delay and phase difference together, we can look at the equivalent phase shift to steer a beam through a given angle so it adds up constructively to create a beam (or lobe) or to cancel destructively to create a null (minimum in the radar beam pattern):

Example 7.3: What is the phase difference required for the output of two sources emitting in phase to produce a time delay of half the wave period (3 significant figures)?

Using the expression: $\dfrac{t}{T} = \dfrac{\phi}{360}$ (**eq 7.5**)

then $\phi = \dfrac{360\,t}{T}$ (**eq 7.6**)

THE ELECTROMAGNETIC SPECTRUM

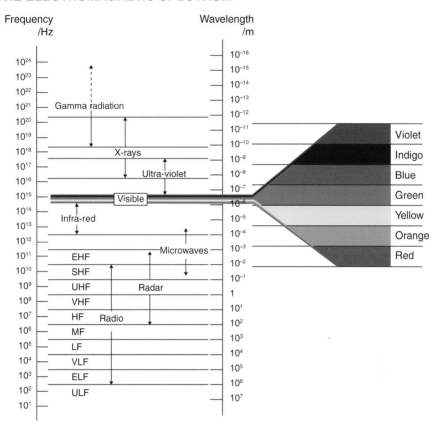

Figure 3.2: *The electromagnetic spectrum.*

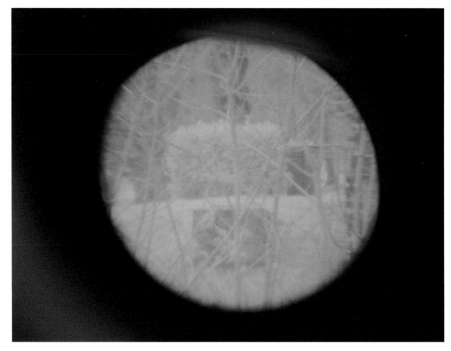

Figure 3.4: *Filtered night vision image display.*

Figure 3.5: *Thermal image of a vessel in port.*

Figure 4.11: *Strong direct optical reflection.*

Figure 4.12: *Weak indirect optical reflection.*

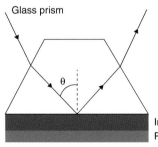

Figure 4.22: *(left) Typical rotation stage reflectivity vs angle experiment for a prism coated with a 40 nm layer of sputtered silver. (right) Layer arrangement: Glass prism coated with indium tin oxide (ITO), a transparent coating, and polyimide (just a few nm) for aligning liquid crystals.*

Figure 4.25: *Mirage observed on a hot UK summer day.*

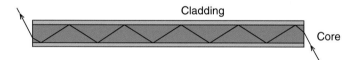

Figure 4.26: *Optical fibre structure and path of a totally internally guided light ray [4.16].*

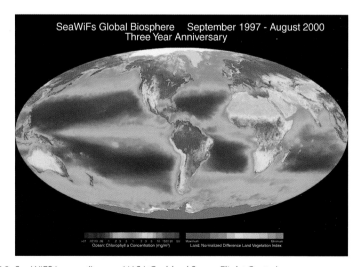

Figure 5.3: *SeaWiFS image. (Image: NASA Goddard Space Flight Center)*

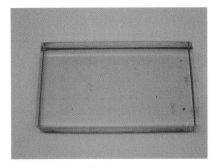

Figure 5.4: *Glass block.*   Figure 5.5: *Glass block on its thicker side.*

Figure 6.3: *Single slit diffraction pattern (note the narrow vertical slit gives rise to a horizontal modulated spreading of the beam).*

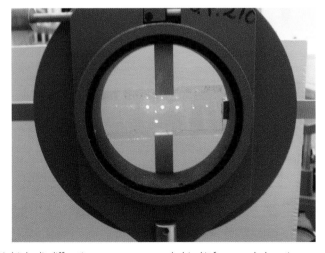

Figure 6.4: *Multiple slit diffraction seen on a screen behind it from a ruled grating.*

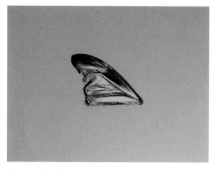

Figure 6.5: *Morpho wing air on natural grating.*

Figure 6.6: *Acetone on natural grating.*

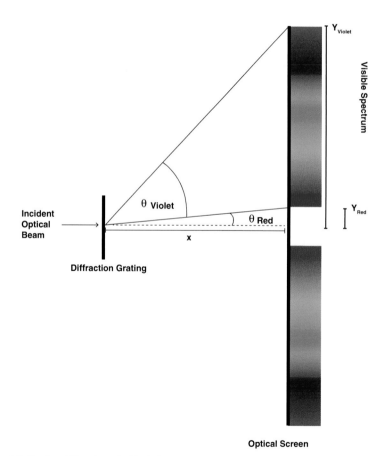

Figure 6.8: *Grating diffraction of white light.*

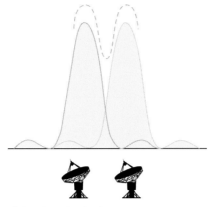

Figure 6.11: *Two telescope dishes detecting in phase.*

Figure 6.12: *The Very Large Array, New Mexico. (Image: Stephen Hanafin)*

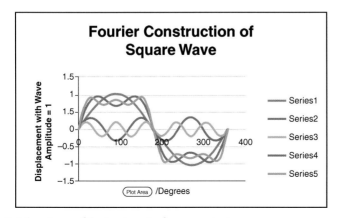

Figure 7.3: *First three terms of the Fourier series for a square wave.*

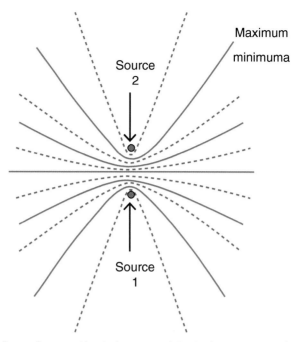

Figure 7.5: *Interference from two identical sources emitting in phase many wavelengths apart.*

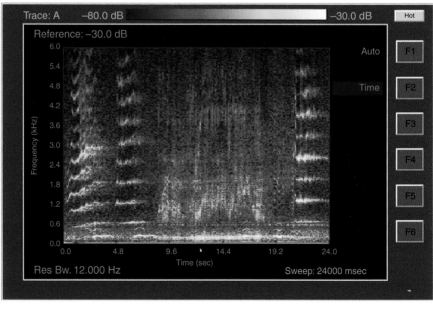

Figure 8.1: *Sonar spectrogram of humpback whale song. (Image: Spyrogumas/Wikimedia Creative Commons)*

so the phase difference will be given by: $\phi = \dfrac{360T}{2T} = 180$ degrees.

The time it takes to steer a beam electronically is very much shorter than the time it takes to mechanically rotate a radar, as all that is required is that the sources start with a very slight time delay with respect to the next adjacent source.

Example 7.4: What is the time delay required to steer a radar beam off the bore sight (the normal straight-through direction) if the required path difference between adjacent sources is 3 cm?

Given that speed = path difference/ time delay $c = \dfrac{path\ difference}{time\ delay}$ **(eq 7.7)**

Therefore: $time\ delay = \dfrac{0.03}{3 \times 10^8} = 1 \times 10^{-10}\,s$, which is an extremely short interval of time

when considered between two adjacent sources.

In a row of 100 sources, as in our earlier example, this will result in an overall time delay from the first to last source of: $99 \times 1 \times 10^{-10} = 99 \times 10^{-10}s-$ which is still much less than the interval of time needed to steer the radar beam mechanically.

7.3 Interference-based navigation-related sensors

Navigation is the indispensable process of planning and controlling vessel movements from one location to another. All navigational techniques involve locating a navigator's position in relation to other known locations, and several important navigation-related sensors help the modern navigator achieve this. Radar navigation uses radar to determine distance from or bearing of objects whose position is known, and is primarily used within radar range of land. Radio navigation, however, uses radio waves to determine position by either radio direction finding systems or hyperbolic systems, such as LORAN-C.

LORAN-C is similar in principle to both Omega and Decca; as of 2010, all three have been discontinued due to the introduction of satellite-based Global Positioning Systems (GPS). Satellite navigation uses artificial Earth satellite systems, such as the American GPS NAVSTAR and Russian GLONASS, to determine position and is used worldwide.

A proposed supercessor version of LORAN-C, e-LORAN, is now gaining ground to relaunch the cancelled radio-navigation network and provide an alternative to the vulnerabilities of having all positional tracking in one GPS basket. However, we will focus briefly on the use of Loran-C as it arises from combinations of two-source interference and there is a great deal we can learn about multiple-source terrestrial radio-navigation for any such future e-LORAN system.

7.4 LORAN-C

LORAN (LOng RAnge Navigation) was a terrestrial navigation system using Low Frequency (LF) radio transmitters and operated using the interference method. Before satellite-based GPS systems, it was the primary location method used in marine applications. The last version of LORAN-C was an ocean navigation aid requiring a special receiver. A master transmitter and slave transmitter emitted pulses at the same frequency, an example of two-source interference. Measuring the difference between pulse arrival times determines on which hyperbolic Line of Position (LOP) relative to the two transmitters a receiver is situated (figure 7.8). In a simplistic case, distance travelled may be measured directly between the twinned transmitters, as for every half wavelength travelled the interference will go from its start value of combined signal strength to a maximum (or minimum) recorded value, to its minimum (or maximum respectively) and then back again, completing one cycle.

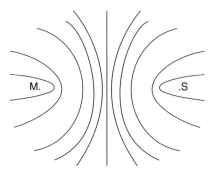

Figure 7.8 *Master and slave transmitter and consequent interference pattern.*

Crossed LOPs from two master-slave pairs using the same master transmitter twice provided an approximate fix in a region with overlapping master-slave pairs.

A real LORAN-C chain consisted of a master transmitter with four separate slave transmitters designated: W, X, Y, and Z. LF transmitters operated at 100 kHz and emitted groups of typically microsecond duration pulses at a specified interval, the Group Repetition Interval (or GRI), unique to each chain and selected to avoid interference with other nearby chains. Each slave station transmission was delayed with respect to the master, the secondary coding delay, to ensure pulse groups were always received in the sequence (M, W, X, Y, Z). The master transmitted 9 pulses while slaves transmitted 8; the master's 9th pulse was used for identification. LORAN was originally an American development of the British GEE radio navigation system used in World War II and had a range of about 400 miles.

In LORAN-C, one station was kept constant in each application, the master, which is paired separately with the other slave stations. Given two secondary stations, the time difference (TD) between the master and first secondary identified one curve, and the time difference between the master and second secondary identified another curve. The intersection of lines was used to determine a geographic point in relation to the position of the three stations. These curves were referred to as Time Delay or TD lines. Essentially, we count the distance we have travelled by counting the number of constructive to destructive half wavelengths moved, as indicated previously. The master station transmitted a series of pulses, and then waited for that amount of time before transmitting the next pulse group sequence. Secondary stations received the pulse signal from the master and waited a preset number of milliseconds, known as the secondary coding delay, before transmitting a response signal. In a given chain, each secondary's coding delay was different, allowing separate identification of each secondary's signal. Modern LORAN receivers displayed latitude and longitude instead of signal time differences, to a high degree of accuracy. Many maritime platforms still have LORAN receivers fitted, uncertain of the future of e-LORAN implementation.

7.4.1 Limitations of LORAN-C

LORAN suffered from the effects of weather and also ionospheric effects close to sunset and sunrise. The most accurate signal is the ground wave that follows Earth's surface along a water path, rather than sky waves (already discussed). Ground waves can actually travel better over water than land, firstly because water surfaces tend to be flat without strong surface scattering effects, and also because water has a higher surface conductivity than land. At night, indirect sky waves, taking paths refracted back to the Earth's surface by the ionosphere, are a problem as multiple signals can arrive from different paths, interfering when we least wish this to occur.

The ionosphere's reaction to sunset and sunrise accounts for disturbances and large errors can occur during these periods. Magnetic storms can also have serious effects as Loran-C is still a radio frequency based system. Due to interference, the very phenomenon LORAN is based upon, and further propagation losses suffered by low frequency signals from both land features and man-made structures, LORAN-C signal accuracy degraded significantly inland. Satellite Global Positioning Systems (GPS) thereby offer significant advantages to land-based users who, of course, make up the majority of modern geospatial customers.

Further detailed discussion of LORAN-C will be found in chapter 2 of Reeds Marine Engineering and Technology Series, Volume 15: *Electronics, Navigation Aids and Radar Theory for Electrotechnology Officers* (ISBN 978–1-4081–7609–2).

7.5 Self-assessment questions

After studying this chapter, you should be able to answer the following questions.

1.   Explain what can happen when two waves of the same frequency meet.

2.   Explain what happens when two sources radiate into the same region of space and sketch the resulting interference pattern.

3.   Sketch the frequency and time domains for a frequency and its first three harmonics. What is the fundamental relationship between frequency and the periodic time T?

4.   If a receiver R a distance x from both sources A and B is moved a distance dx towards source A, show with the aid of a diagram how you can calculate the distance needed to move from the central maximum to the second destructive minimum. What is the expression for equivalent time delay, path difference, and phase difference?

5.   What will be the result of two sources emitting exactly out of phase for waves arriving at a point in space midway between the two sources?

6.   For two waves of the same frequency emitting with a time delay of T/4 and a physical path distance of $\lambda/3$ and a phase difference of 180 degrees, what will be the resultant of the two waves arriving at the same point in space?

7.   Find the time delay required in terms of the periodic time T if the path difference between two adjacent sources is 2 cm and the wavelength is 10 cm.

8.   Consider a regularly spaced square array of sources in both the vertical and the horizontal direction. If there are 900 sources in total and the sources in every row and column are spaced by half a wavelength, what will be the minimum resulting beam width if the wavelength is 2 cm (2 decimal places)?

9.   Find the time delay required to steer a radar beam off the bore sight (the normal straight-through direction) if the required path difference between adjacent sources is 2 cm.

10.   Explain how LORAN-C operates.

Acoustic Waves, the Doppler Effect and Decibels

'It would be possible to describe everything scientifically, but it would make no sense; it would be without meaning, as if you described a Beethoven symphony as a variation of wave pressure.' Albert Einstein

8.1 Acoustic pressure waves

In figure 2.2 we saw the gradual build-up of a longitudinal or sound wave disturbance. In such a wave there are local regions in the material where particles are closer together (the density and local pressure are rising) and other regions where particles are moving apart (the density and local pressure are falling). Such waves propagate through a material by fluctuations in pressure and as such they are often referred to as pressure waves. Sound in air and sonar waves are of this particular type of wave. Sensing systems under water will use sonar rather than electromagnetic waves, as although sonar waves travel much more slowly, they have much lower loss than electromagnetic waves. Typical ranges of sound in the Earth's seas are typically of the order of tens of kilometres, but, if conditions are right, some sound waves can be detected across an entire ocean.

8.2 The use of sound in the sea

Practical uses of sound in the oceans can be divided into two very different types of sensing system. The first group is that of sensors that transmit sound through water, which is usually detected by the same sensor system (*active*), and the other group in which sound, already present in the sea, originates from other sources, and is detected only (*passive*). Sound use is commonly referred to as SONAR, which stands for SOund Navigation And Ranging. Sonar is a technique that uses sound propagation (usually underwater) to navigate and to detect other vessels or communicate beneath the water surface. Sonar has many useful applications, which can be divided into two groups, namely that of civilian and military applications, as can be seen in Tables 8.1 and 8.2 respectively.

Acoustic releases	These disengage buoyed equipment used at depth when suitably instructed.
Beacons	These emit powerful sound signals as an aid to navigation.
Communications	Data can be transmitted by acoustic modulation of sound waves in a similar way to that in optical modulation of a light beam, but with narrower available bandwidth.
Doppler navigation	Angled beams generate a Doppler shift due to platform movement relative to the stationary seabed.
Echo sounder	These emit short pulses vertically down and times the return of the echo in a similar way to radar with electromagnetic radiation. These systems operate at high frequency and have a narrow beam.
Fish echo location	Trawler fleets use forward-looking active sonar to locate schools of fish.
Side-scan sonar	Side-scan sonar has wide beams angled to the vertical surveying large widths of seabed either side of the ship's track.
Sub-bottom profiler	These use a high power, low frequency source to produce sound waves that partially penetrate the seabed.
Transponders	These transmit appropriate signals when interrogated by a suitable signal, similar to aviation Identification Friend or Foe (IFF) systems.
Underwater floats	These emit an identifiable signal so they can be tracked over time to measure ocean currents.

Table 8.1: *Civilian applications of sonar.*

Acoustic mine	An underwater mine can detect general levels of ship's noise. The more complex modern sensors can detect the spectral content of noise, allowing precise identification of underwater targets.
Homing torpedoes	Active and passive high frequency sonar can be used for this type of application.
Minehunter	These use very high frequency active sonar. These types of sonar require high definition but can only operate at short range due to the high losses at these frequencies for the same reasons encountered with radar systems.
Sonobuoys	These are passive listening devices that transmit data by radio from the surface.
Submarine and surface ship detection	Submarine and surface ship detection utilise active and passive sonar operating at low and medium frequencies. They are used to achieve long range detection but have poor definition when compared with minehunter HF systems.
Underwater telephone	Voice modulation can be operated on relatively low frequencies and typically can operate on an 8 kHz carrier.

Table 8.2: *Military applications of sonar.*

8.3 Historic background of sonar

8.3.1 The sonar spectrum

The frequency range of sonar longitudinal waves underwater is known as the sonar spectrum, and for sound waves is equivalent to the electromagnetic spectrum we have already discussed for electromagnetic waves. This sonar can be displayed using a false coloured spectrograph (such as that shown in figure 8.1, see plate section).

8.3.2 History of sonar

Sonar can help measure the echo characteristics of 'targets' in water. The acoustic frequencies used in sonar systems vary from very low frequencies (infrasonic) to extremely high frequencies (ultrasonic). The study of underwater sound is often referred to as hydroacoustics or underwater acoustics. Although some animals – bats and cetaceans (whales and dolphins) – use sound for communication and object detection extensively, the first recorded use by humans in water is believed to be by the Italian Leonardo da Vinci (1452–1519) in 1490. He was reputed to have inserted a tube into water to detect vessels by placing his ear to the tube. Little changed until the 19th century, when an underwater bell was often used to provide warning of hazards.

The modern use of sound to actually *echo locate* underwater in a similar way to radar in the above water environment was prompted by the sinking of the *Titanic* in 1912. The world's first patent for an underwater echo ranging device was made by English meteorologist Lewis Richardson just one month after the *Titanic* disaster, while a German physicist, Alexander Behm, obtained an echo sounder patent soon after, in 1913.

A Canadian engineer, Reginald Fessenden, while working for the Boston Submarine Signal Company, built a first experimental system in 1912, which was fully tested in 1914 off the Grand Banks of Newfoundland, Canada. In this test, Fessenden demonstrated depth sounding and underwater communications using Morse code, as well as echo ranging (detecting an iceberg at 2 miles range). During World War I, the urgent need to detect submarines redoubled research into the use of sound. The British made early use of underwater listening devices (hydrophones), while in 1915 a French physicist, Paul Langevin, worked to develop active sound devices for submarine detection.

Side-scan sonar is a category of sonar used to create an image of large areas of the sea floor. It can be used to conduct underwater surveys and is often used in conjunction with sea floor sampling to provide an understanding of the differences in material and texture type of the seabed. Side-scan sonar imagery is a commonly used tool to detect wreckage items and other sea floor obstructions that may be hazardous to shipping or to sea floor installations used by the gas and oil industry. The status of seabed pipelines and cables can safely be conducted using side-scan sonar. Side-scan data are often acquired beside ocean depth (bathymetric) soundings and with sub-bottom profiler data can provide pictures of the seabed's shallow structure. Side-scan sonar is also used in fishing research, dredging operations and environmental studies, and in military applications such as mine detection.

Side-scan uses a sonar device that emits fan-shaped or conical pulses towards the sea floor across a wide angle range perpendicular to the sensor's path through the water, and may be towed by a submarine or surface vessel, mounted on the ship's hull. The intensity of acoustic reflections from the sea floor is recorded in a series of cross-track slices, which are then 'stitched' together in the direction of motion. These slices form an image of the sea floor within the beam's coverage width (or swathe). Sound frequencies used in side-scan sonar typically ranges between 100 and 500 kHz; higher frequencies provide enhanced resolution but losses at these higher frequencies result in significantly reduced sonar range.

Figure 8.2: *Side-scan sonar array image of a shipwreck. (Image: Australian Transport Safety Bureau)*

8.4 The Doppler effect

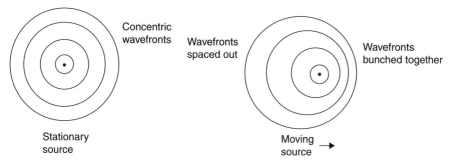

Figure 8.3: *The Doppler effect, showing the change on the wave fronts due to the moving source.*

The *Doppler effect* (or *Doppler shift*) is the change in wave frequency observed relative to a moving wave source, and is named after the Austrian physicist Christian Döppler (1803–1853), who proposed it in 1842 while in Prague. It is often heard when a vehicle sounding its siren approaches, passes, and then recedes from an observer. Compared to the emitted frequency, the received frequency is higher during the approach, identical to the emitted frequency at the instant of passing by, and relatively lower as it moves away. The frequency shift observed depends upon the relative velocity of the motion that takes place between an observer and the source of waves. At the instant when an object moves at 90 degrees to the observer, it will appear to emit at the same frequency as that transmitted, i.e. no Doppler shift is observed. Extensive use of the Doppler effect is made by both modern radar and sonar systems for electromagnetic and acoustic waves respectively.

The Doppler effect applies both to acoustic waves, such as sound in air and sonar waves in water, and to electromagnetic waves. For example, when the source of radar waves moves towards an observer, successive wave crests will be emitted from a position a little closer to the observer than the one before. Hence, each wave takes slightly less time to reach the observer than those previously. The time between the arrival of successive wave crests at the observer is thus reduced, resulting in an increase in the recorded frequency. While they are travelling, the distance between successive wave fronts is reduced so the waves 'bunch together'. In a similar way, if the source of waves moves away from the observer, each successive wave will be emitted from a position slightly further away from the observer than the previous one (figure 8.3), so the arrival time between successive waves is thus increased, reducing the recorded frequency. The distance between successive wave fronts is then increased, so the waves 'spread out'.

For waves that propagate in a medium, such as sound waves travelling in air or sonar waves travelling underwater, the velocity of an observer and of the source are relative to the medium in which the waves are transmitted. The total Doppler effect may thus result from motion of the source, or motion of the observer, or relative motion of the medium itself. As long as there is *relative motion* taking place, a Doppler shift is observed. For waves that do not require a medium, such as light or other electromagnetic waves such as radar, only the relative difference in velocity between the observer and the source should be considered.

8.5 Relative velocity

Consider a source of waves travelling at velocity Vs in the direction indicated in figure 8.4. A target will return Doppler shifted echoes if travelling at velocity Vt in the direction shown. By convention, the sign of the velocity is positive in the direction of the source waves travelling out towards the target. The relative velocity is recorded as the difference of the velocity once the correct sign convention has been applied. Further details about marine Echo Sounders are to be found in chapter 2 of Reeds Marine Engineering and Technology Series, Volume 15: *Electronics, Navigation Aids and Radar Theory for Electrotechnology Officers* (ISBN: 978–1-4081–7609–2).

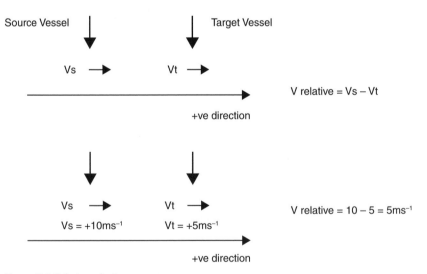

Figure 8.4: *Relative velocity.*

An example of the general case is also given in figure 8.4 above. The source is moving right at a velocity of 10 metres per second, while the target moves in the same direction at a velocity of 5 metres per second. Clearly the source will 'catch up' and eventually overtake the target at some point, because it is travelling faster. The relative velocity is found by taking the difference in the two velocities in the same positive direction.

So: $V_{relative} = V_{source} - V_{target}$ **(eq 8.1)**

8.6 Doppler effect: A moving source

Let us consider a source of electromagnetic waves travelling at speed c. What happens when the source of waves moves towards you, a stationary observer? You will encounter more waves per unit time than you did before, so the frequency will be increased relative to a source at rest. The shift occurs because the wavelength has been lowered by the movement of the source towards you, as the waves bunch together.

Let the speed of the source of waves, the aircraft or ship, moving towards you have a value of $V_{relative}$. When nothing moves, the wave front moves a distance ct, where t is the period of time we are considering and c is the speed of the waves, so the number of wavelengths detected or received in time $t = \dfrac{ct}{\lambda}$, and the rate the

receiver detects wave fronts over the time t (which of course is the frequency) $= \dfrac{ct}{\lambda t} = \dfrac{c}{\lambda}$, as expected. However, when the source moves at speed $V_{relative}$, the number of wavelengths in that extra distance covered $(c + V_{relative})t$ will mean an increase in the number of wavelengths intercepted, i.e. $\dfrac{(c + V_{relative})t}{\lambda}$ and the rate at which the receiver detects wavelengths is the detected but shifted frequency:

$$f_{received} = \frac{\frac{(c + V_{relative})t}{\lambda}}{t} = \frac{(c + V_{relative})}{\lambda} \quad \text{now since } \lambda = \frac{c}{f_{transmitted}}$$

$$f_{received} = \frac{(c + V_{relative})f_{transmitted}}{c}$$

But $f_{received} = f_{transmitted} + \Delta f$ where Δf is the frequency shift.

So $(f_{transmitted} + \Delta f)c = (c + V_{relative})f_{transmitted}$

And thus: $cf_{transmitted} + \Delta fc = cf_{transmitted} + V_{relative}f_{transmitted}$

Simplifying terms for one way transmission:

$\Delta fc = V_{relative}f_{transmitted}$

So that: $\Delta f = \dfrac{V_{relative}}{c}f_{transmitted}$ **(eq 8.2)**

When the relative velocity is closing with the observer and target approaching (negative in our convention), the frequency shift will be positive, but if the relative velocity is opening with the observer and target separating (positive in our convention), the frequency shift will be negative, i.e. to lower than the transmitted frequency.

More generally for all sorts of waves travelling in a medium at speed V:

$\Delta f = \dfrac{V_{relative}}{V}f_{transmitted}$ **(eq 8.3)**

Note that for an echo the target becomes a secondary source of radar waves, which will Doppler echo shift again (twice in total) back to the wave detector.

$\Delta f \; echoes = \dfrac{2V_{relative}}{V}f_{transmitted}$ **(eq 8.4)**

Example 8.1: If the transmitted radar frequency is 10 GHz, the relative velocity is 10 metres per second and the speed of light c is taken as usual, find the echo frequency shift observed (1 decimal place).

The value of the Doppler frequency shift will be found by taking the above equation and substituting for the given values:

$\Delta f = 2 \times 10 \times \dfrac{10^9}{3 \times 10^8} = 66.7 \; metres \; per \; second$

The frequency of the sound that a source emits of course does not change. To understand what happens, consider the following analogy: a ball is thrown every second in a man's direction. Assume that consecutive balls travel with constant velocity. If the thrower is stationary, the man will receive one ball every second at a regular repeatable rate. However, if the thrower moves towards the man, the man will catch balls more frequently because the balls will be less spaced out. The inverse is true if the thrower moves away from the man. So it is the wavelength that is affected; as a consequence, the received frequency is also affected. The velocity of the wave remains constant while wavelength changes; thus the frequency also changes. In the ball analogy, the speed of the balls depends on the speeds of the thrower and receiver, which is not the case of the speed of light wave front velocity, which remains constant.

When a vehicle passes an observer, the radial velocity component between the source and the target will fall to zero at some point. Hence, to consider off direct axis situations – a more complicated scenario – means obtaining the relative velocity by considering the off axis angles to get the source and target velocity components along the direct line between them. Hence:

$$V_{relative} = V_{source}\cos\theta_{source} - V_{target}\cos\theta_{target} \qquad \textbf{(eq 8.5)}$$

which is the relative velocity component along the direct line between the source and target (figure 8.5).

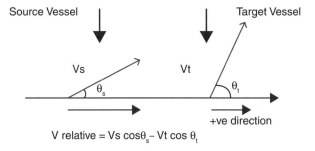

Figure 8.5: *Off axis Doppler consideration.*

The Doppler effect is used in astronomy to evaluate the frequency shift of spectral lines. The Doppler effect for electromagnetic waves such as light is of great use in astronomy and results in either a so-called red shift or blue shift. It is used to measure the speed with which stars and galaxies are approaching or receding from us to obtain the radial velocity.

8.7 Doppler radar

The Doppler effect is used in some types of radar to measure the velocity of detected objects. A radar beam can be transmitted towards a moving target, e.g. by a policeman using a speed camera pointed at a motor car to check its speed. As the waves approach or recede from the radar source, each successive radar wave has to travel further to reach the car before being reflected and detected at the receiver (collocated with the source). As each wave has to move further, the gap between successive waves increases, increasing wavelength. Calculations from the Doppler effect can accurately determine a car's velocity, taking into account wind speed and its direction relative to the car.

Other applications include medical blood flow imaging and flow measurement generally, with instruments such as the Laser Doppler Velocimeter (LDV), and the Acoustic Doppler Velocimeter (ADV) developed to measure velocities in fluid flow. The LDV emits a light beam and the ADV emits an ultrasonic pulse to measure the Doppler shift in reflected wavelengths from particles moving with the flow. Flow is computed as a function of the water velocity and phase. This technique allows non-intrusive flow measurements, with high precision and frequency.

In satellite communications, fast moving satellites can have a Doppler shift of several tens of kHz relative to a stationary ground station. The speed, and thus the magnitude of the Doppler effect, changes due to Earth curvature. Dynamic Doppler compensation, where the frequency of a signal is changed multiple times during transmission, is used so the satellite receives a constant frequency signal. In this way, one now obsolescent military satellite system, Transit, used the Doppler shift to provide a degree of navigational positional fixing prior to the advent of GPS.

8.8 Decibels

With acoustic signals, the variable range of signal strengths is an important factor to consider. Intensity and power ratios can sometimes be extremely large and yet on the other hand they can at times be extremely small. For this reason, they are often better expressed using *decibels* (dB). The decibel is a logarithmic unit named after Alexander Graham Bell (1847–1922), the accredited Scottish inventor of the telephone, and a power ratio can be converted to decibels by the relation:

Power ratio in decibels = 10 log (P_2/P_1) (**eq 8.6**)

where P_1 is the initial power and P_2 is the final power.

Example 8.2: If the initial power level P_1 is 5 W and the final power level P_2 is 100 W, what is the power ratio in decibels (2 decimal places)?

Power ratio in decibels = 10 log (P_2/P_1) = 10 log (100/5)

$= 10$ log $20 = 13.01$

The decibel originates from methods used to quantify reductions in audio levels in telephone circuits. Such losses were originally measured in units of *Miles of Standard Cable* (MSC), where 1 MSC corresponded to the power loss over 1 mile of standard telephone cable at a frequency of 795.8 Hz, and matched the smallest attenuation detectable to the average listener. The *Transmission Unit* (TU) was devised by engineers of the Bell Telephone Laboratories in the 1920s to replace the MSC. 1 TU was defined as ten times the base-10 logarithm of the ratio of measured power to a reference power level. The definitions were conveniently chosen so that 1 TU approximately equalled 1 MSC (specifically, 1.056 TU = 1 MSC). In 1928, the Bell system renamed the TU to the decibel. Along with the decibel, the Bell System defined the *bel*, the base-10 logarithm of the power ratio, in honour of their founder and telecommunications pioneer Alexander Graham Bell. The bel, $\log (P_2/ P_1)$, is rarely used, as the decibel is a much more useful working unit.

Since intensity is power per unit area, the above equation can also be written using intensity instead of power. It is useful to remember that a positive decibel value means an increase in power or intensity (gain) and a negative decibel value a decrease (loss). Useful decibel values are +3 dB, which multiplies the power or intensity by a factor of 2, and −3 dB, which halves the power or intensity. Further values are given in Table 8.3. Most of the table can be generated relatively quickly with only the 3dB and −7dB power ratios defined (this is left to the astute reader to prove).

dB	Power ratios P_2/ P_1
0	1
1	1.26
2	1.6
3	2
4	2.5
5	3.2
6	4
7	5
8	6.3
9	8

Table 8.3 *Basic power ratios.*

Engineers make extensive use of decibels as they can represent both gain and loss. In any signal processing system there will be amplifying components (gain) providing an increase in signal level, as well as interconnections and long cables where there is loss or reduction in signal level, managing very large and very small quantities in a convenient dimensionless manner.

Using decibels enables simple addition and subtraction to find the overall gain or loss of many engineering systems.

8.9 Acoustic waves

Now we have introduced the decibel we are in a place to talk about acoustic units and, by returning to our original SHM concept of wave motion, to look at velocity and peak velocity of the displaced wave.

The intensity of sound in water depends on the square of sound pressure. The standard unit of intensity is that of a field with a Root Mean Square (RMS) sound pressure of 1 Pascal. Observed sound intensities are usually described as plus or minus so many decibels with respect to 1 Pascal RMS.

A standard source level is such that it gives rise to a sound intensity of 1 Pascal RMS at a distance of 1m in the direction of greatest intensity. Thus we can describe any source level S in decibels with respect to the standard source. This is the same as dealing with intensity of sources generally, already covered in Chapter 5.

In a large homogeneous body of water that has the same properties in every direction, it is observed that energy spreads out uniformly so that the intensity at a range R from a source level S in decibels would be, in the absence of losses:

$I = S - 20 \ log \ R$ in terms of logs to base 10   **(eq 8.7)**

Now, this arises from consideration of a source radiating isotropically (be that air or water) according to the following relationship:

$$I' = \frac{P}{4\pi \ R^2}$$

where P is the radiated Power and R the range.

Rearranged slightly:

$$I' = \frac{S'}{4\pi \ R^2}$$

Taking logs:

$\log I' = \log S' + \log R^{-2}$ and multiplying by factor of 10 to put the range factor into decibels:

$$10\log I' = 10\log S' - 20\log R$$

Which, with a little sleight of hand, allows us to redefine I' and S' (with $I = 10\log I'$ and $S = 10\log S'$, remembering that the source S is in decibels anyway) so that:

$$I = S - 20\log R$$

There is, however, further attenuation due to the conversion of sound energy to heat within the medium, which, if reasonably considered to be a linear process with distance, can be considered to have a value of a decibels per metre. So a more correct expression found for sound intensity in water at the target will become:

$$IT = S - 20\log R - aR \qquad \textbf{(eq 8.8)}$$

This expression is valid for uniform spreading in a large water volume and quite accurate for sonar echo sounding, but will be in error where energy is transmitted in a horizontal direction where depth is small compared with the range. In this case, spreading of wave energy will be severely limited by the water boundaries above and below (so multiple reflections occur from the bottom) or where refraction is significant, causing sound to be channelled, and so greater energy will be found in some directions and less in others, very different to that which might have been expected from uniform isotropic spreading. The last term to be added to our simple model is the target level, which is defined in decibels relative to the standard. Incident acoustic energy striking a target will cause the target to act as a secondary source so that it in turn acts as a standard source (basically, it is a perfect reflector with no loss), so the target will be a 0 dB target (as Signal/Reference = 1).

If the range of the target from the sonar system is R metres, the re-radiated target level is T and the intensity incident upon the target is IT, then the intensity of the sonar echo signal at the receiver IR will be as follows:

$$IR = IT + T - 20\ \log\ R - aR \qquad \textbf{(eq 8.9)}$$

However, we know that: $IT = S - 20\ \log\ R - aR$

Combining these two equations allows us to obtain the fundamental sonar equation:

$$IR = S - 40logR - 2aR + T \qquad \textbf{(eq 8.10)}$$

This equation applies to small targets that subtend a small angle to the sonar system much less than the beam width, i.e. they are considered to be point sources and not distributed targets. It refers only to targets on the direct axis of the transmitted sonar beam, as we have defined source level by reference to the direction of maximum intensity, which unless a beam is steered off axis will occur in this direction. Off-axis conditions have already been considered. Where a distributed target occurs (such as with the seabed), much larger than compared with the sonar system beam width, a larger target cross section will become involved in reflection as the range increases.

Consider the seabed, which will tend to fill the sonar beam width irrespective of the range of any specific point on the seabed (you can always see it, wherever you are). The seabed thus gives rise to an echo intensity at the sonar receiver that varies with range as $-40logR - 2a\,R$ and whose accurate prediction depends on the nature of the seabed and the directional properties of the transmitter as well as its source level.

For a complete study, it is important to know the target strength per square metre for the surface considered, how it varies with incident angle, and how the radiated energy of the source varies with angle. As this is not relevant to our brief introduction to acoustic waves, it will not be considered further here.

However, in terms of considering pressure, particle velocity and amplitude, let us consider the simple case of a *plane wave* proceeding in the positive x direction. We will assume, as we have done previously, a progressive cosinusoidal pressure wave of the form:

$$p = Pcos(\omega t - kx) \qquad \textbf{(eq 8.11)}$$

where p is the instantaneous pressure, P the peak pressure, and c is here the phase velocity of the wave motion as discussed in Chapter 2.

To obtain the velocity u, the peak velocity U and the amplitude of the resulting displacement, consider the following.

Take a tube of unit cross-sectional area parallel to the horizontal x-axis. Any element of thickness dx will be subject to a force of $\left(\dfrac{\partial p}{\partial x}\right) dx$ and has a mass of ρdx. Thus the acceleration ($F = ma$) acting upon the element of thickness dx is:

$$\left(\frac{1}{\rho}\right)\left(\frac{\partial p}{\partial x}\right)$$

Acceleration is also the velocity gradient, so:

$$\frac{du}{dt} = \left(\frac{1}{\rho}\right)\left(\frac{\partial p}{\partial x}\right)$$

Which by integrating can be given as: $u = \left(\dfrac{1}{\rho}\right) \displaystyle\int \left(\dfrac{\partial p}{\partial x}\right) dt$ **(eq 8.12)**

Using **(eq 8.11)**

$$p = P\cos(\omega t - kx)$$

So $\dfrac{\partial p}{\partial x} = -Pk\, \sin(\omega t - kx)$ **(eq 8.13)**

Hence by substitution of 8.13 into 8.12 then:

$$u = \left(\frac{1}{\rho}\right)\int -Pk\,\sin(\omega t - kx)\, dt$$

And $u = \left(\dfrac{-Pk}{\rho}\right)\displaystyle\int \sin(\omega t - kx)\, dt$

Or $u = \left(\dfrac{-Pk}{\rho}\right)\left(-\dfrac{1}{\omega}\right)\cos(\omega t - kx)$

And since $\omega = ck$

$$u = \left(\frac{-Pk}{\rho}\right)\left(-\frac{1}{ck}\right)\cos(\omega t - kx)$$

Hence: $u = \left(\frac{P}{\rho c}\right)\cos(\omega t - kx)$ **(eq 8.14)**

Now, the peak pressure and the peak velocity are shown to be related by: $U = \left(\frac{P}{\rho c}\right)$ as the maximum value of $\cos(\omega t - kx)$ is 1.

So to find the amplitude of individual particle vibrations for any point in the positive x direction:

$$u = U\cos(\omega t - kx)$$ **(eq 8.15)**

The amplitude is given by: $A = \int_{\omega t=0}^{\omega t=\frac{\pi}{2}} u\, dt$ for a *quarter cycle* from maximum velocity to zero velocity.

This choice of limits is important, as trying to integrate over the full 2π limits will result in a net amplitude of zero. (If in doubt, think about your respective position compared to the old one after a complete cycle!) This is left to the reader to consider.

Now $A = \int_{\omega t=0}^{\omega t=\frac{\pi}{2}} U\cos(\omega t - kx)\, dt$

$A = U[\sin(\omega t - kx)]_0^{\frac{\pi}{2}}$

$A = \frac{U}{\omega}\left[\sin\left(\frac{\pi}{2} - kx\right) - \sin(0 - kx)\right]$

$= \frac{U}{\omega}\left[\sin\left(\frac{\pi}{2}\right) - \sin(0)\right]$

$$= \frac{U}{\omega}[(1) - (0)]$$

So that: $A = \dfrac{U}{\omega}$

And using our earlier expression for U, ie $U = \left(\dfrac{P}{\rho c}\right)$:

Thus: $A = \left(\dfrac{P}{\omega \rho c}\right)$

The quantity ρc is called the *characteristic acoustic impedance* of the medium. The acoustic impedance is similar to its electrical counterpart, with pressure analogous to voltage and velocity analogous to electrical current.

In military applications, the Doppler shift of a sonar target is used to obtain a submarine's speed using both passive and active sonar systems. As a submarine passes a passive sonobuoy, there will be a Doppler shift, and the speed and range from the sonobuoy can be calculated. If the sonar system is mounted on a moving ship or another submarine, then the relative velocity can be calculated.

8.10 Ultrasound

Ultrasound or ultrasonic waves are sound waves with frequencies above about 20 kHz and are undetectable by humans as they are outside the audible range. Ultrasonic waves thus have very small wavelengths and can be transmitted as a beam with very little difficulty – making them useful for several applications. Echo sounding is used to detect flaws in metal and as ultrasound for pre-natal screening of pregnant women. Ultrasonic waves partly reflected from boundaries between two different media, so ultrasonic waves reflect well at the boundaries between soft tissue and air or body tissue. Ultrasound has several advantages over other methods, especially for non-destructive testing of vessel hull layer defects, namely they are non-invasive, low cost and provide real or at least near real-time imagery.

8.11 Self-assessment questions

After studying this chapter, you should be able to answer the following questions.

1.   What is a decibel and where it is used?

2.   Prove the equation used to calculate a one-way Doppler shifted signal due to relative motion between a source of waves and an observer.

3.   If the transmitted radar frequency is 10 GHz and the relative velocity is 10 metres per second and the speed of light c is taken as usual, find the one-way frequency shift observed (1 decimal place).

4.   In the case of a Doppler shifted echo, calculate the velocity of the target if the observer is stationary, the transmitted frequency is 9.54 GHz and the echo shift is 3 kHz (3 significant figures).

5.   Use the following expression for relative velocity:

$V_{relative} = V_{source}\cos\theta_{source} - V_{target}\cos\theta_{target}$ to calculate the echo Doppler shift if the source has a value of 10 metres per second on a bearing of 30 degrees and the target has a value of 25 metres per second on a bearing of 120 degrees and the radar is transmitting on a frequency of 3.5GHz.

6.   An amplifier has a power gain of +35 dB. If an input power is 38 mW, what is the output signal power (3 significant figures)?

7.   If the initial power level P_1 is 5 W and the final power level P_2 is 100 W, what is the power ratio in decibels (2 decimal places)?

8.   By using the following expression for the instantaneous pressure $p = P\cos(\omega t - kx)$, obtain the velocity u, the peak velocity U and the amplitude of the resulting displacement.

9.   Compare sonar with electromagnetic waves for underwater applications.

10.   Compare the civilian and military applications of modern sonar.

Basic Mathematical and Calculator Operations

'I think there is a world market for maybe five computers.' Thomas Watson, Chairman of IBM, 1943

Calculator computation operations is a subject area that is often neglected. It is assumed that students know how to undertake quite difficult arithmetical operations and operate a programmable calculator flawlessly at the first attempt. However, this is often not the case. We hope to provide a thorough beginners' guide to the subject here, considering as example a commonly available CASIO series calculator. By the end of this chapter you should have a much firmer grasp of the essential maths and calculator principles necessary for all maritime engineering and physics courses you may undertake and be able to apply your understanding of maths and calculator use to conduct subject-relevant calculations. You should also leave with an improved understanding of how accurate computer-based calculations can help you develop your own important maritime skills (e.g. accurate determination of relative motion, etc.).

9.1 Basic mathematical operations

It is assumed that all students can add, subtract, multiply and divide manually and that they will master the calculator operations demonstrated in this chapter. Students or maritime 'officers under training' should be able to work with numbers in *standard form*, i.e. unit, decimal and a power of ten.

For example, 22675000 is expressed as 2.2675×10^7, and 0.0000022675 as 2.2675×10^{-6}.

Students should be able to use standard multipliers when necessary, eg 2.377×10^7 J can be expressed as 23.77 MJ and 4.679×10^{-5} m as 0.4679 μm.

The following basic mathematical operations are used during the course:

9.1.1 Expansion of brackets

(i) $(a + b + c + d) = a + b + c + d$

(ii) $a \times (b + c) = ab + ac$

(iii) $(a + b) \times (c + d) = ac + ad + bc + bd$

(iv) $(a + b) \times (a + b) = a^2 + ab + ab + b^2 = a^2 + 2ab + b^2$

(v) $(a + b)^4 = a^4 + 2a^3b + a^2b^2 + 2a^3b + 4a^2b^2 + 2ab^3 + a^2b^2 + 2ab^3 + b^4 = a^4 + 4a^3b + 6a^2b^2 + 4ab^3 + b^4$

Note: $ab = ba$. ba will be written by convention with the lowest letter in the alphabet given first, thus ba becomes ab.

9.1.2 Manipulation of indices

(i) $a^n \times a^m = a^{(n+m)}$ e.g. $a^3 \times a^5 = a^8$

(ii) $a^n / a^m = a^{(n-m)}$ e.g. $a^3 / a^5 = a^{-2}$

(iii) $a^{n(m)} = a^{nm}$ e.g. $a^{3(5)} = a^{15}$

(iv) $a^{-n} \times a^m = a^{m/n}$ e.g. $a^{-3} \times a^5 = a^{5/3} = a^{1.67}$

Further discussion of logarithms is to be found in chapter 1 of Reeds Marine Engineering and Technology Series, Volume 1: *Mathematics for Marine Engineers* (ISBN 978–1-4081–7556–9).

9.1.3 Rearrangement of equations

When moving an added (or subtracted) term from one side of an equation to the other, the rule is: subtract it from (or add it to) both sides so that it cancels on its original side, e.g.:

(i) Suppose $a = b + c$, then to move c: $a - c = b + c - c$ so: $a - c = b$

(ii) Suppose $a = b - c$, then to move c: $a + c = b - c + c$ so: $a + c = b$

(iii) Suppose $a + b = c$, then to move b: $a + b - b = c - b$ so: $a = c - b$

When moving a multiplied (or divided) term from one side of an equation to the other, you must divide (or multiply) both sides by the same term, so that it cancels on its original side. If the term you want is then left on the bottom, invert both sides of the equation.

(iii) Suppose $a = b \times c$, then to move b: $a / b = b \times c / b$ so: $a / b = c$

Remember, whatever you do to one side of an equation you must do exactly the same to the other side to 'balance the scales'.

Complex equations can be made simpler by bracketing together terms until the bracketed terms resemble a more simple equation.

For example: $A = 3b + 4d + 5c - 6e$

so: $A = B + C$ where $B = (3b + 4d)$ and $C = (5c - 6e)$

This technique is very useful when using a calculator to solve complex equations.

9.2 Graphs and graphical representation of experimental data

There are a number of important rules that must be obeyed when drawing and using graphs.

(i) Both the vertical and horizontal axes should be labelled with the name of the variable and the appropriate units, e.g. a graph of transmitted radar beam intensity against angle should be labelled as shown in figure 9.1.

(ii) The graph must have a title, as indicated.

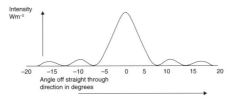

Figure 9.1: A graph of transmitted radar beam intensity vs. off-axis angle.

(iii) Where data is available, the axes should have numerical values indicated. Care must be taken when using graphs where the origin does not represent zero of one, or both, of the variables as graphs can easily be misrepresented (look at the graphs used in newspapers, which often deliberately do this to exaggerate the point they are trying to make) or are simply misinterpreted.

(iv) When plotting a graph where data is obtained by measurement and not by calculation, you should not join up all the points but should instead draw a graph that shows the *general behaviour* or *trend* of the data.

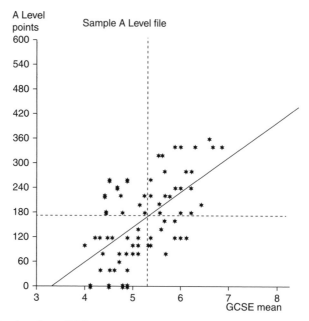

Figure 9.2: *A Level results vs. GCSE mean.*

Further discussion of graphs is to be found in chapter 6 of Reeds Marine Engineering and Technology Series, Volume 1: *Mathematics for Marine Engineers* (ISBN 978–1-4081–7556–9).

9.3 Logarithms

Any number 'X' can be expressed as another number 'a' raised to some power 'n',

i.e. $x = a^n$ where 'n' is called the logarithm of 'x' to the base 'a', eg $100 = 10^2$

The following equation can be written: $\log_a x = n$

'x' can be found from the above equation by taking the inverse log of 'n',

i.e. inverse log n = inverse log $(\log_a x) = x$ (inverse log is called the *antilog*)

Although we now use calculators to compute the values of products and ratios, logarithms provide a convenient and powerful method of carrying out the same computations using only the processes of addition and subtraction – which is very useful without a calculator!

The simple rules of indices already covered show how this can be done.

If $x = a^n$ and $y = a^m$ then $xy = a^n \times a^m$ and so $xy = a^{n+m}$

Now writing these as logarithms gives: $n = \log_a x$

$$m = \log_a y$$

$$n+m = \log_a (xy)$$

i.e. $\log_a (xy) = \log_a x + \log_a y$ the first rule of logarithms

and inverse log $(\log_a (xy)) = xy =$ inverse log $(\log_a x + \log_a y)$

The important rules of logarithms are listed below:

$\log_a (xy) \qquad = \log_a x + \log_a y$

$\log_a (x/y) \qquad = \log_a x - \log_a y$

$\log_a (x^n) \qquad = n (\log_a x)$

inverse log $(\log_a x) = x$

Most calculators have two types of logarithms. The one labelled *log* has the base 'a' = 10 and is called the *logarithm to base 10*. The one labelled *ln* has the base 'a' = e where 'e' has the value 2.718… (an important value in science) and is called the *natural logarithm*. This is discussed further in Appendix D. It is important when you use your calculator and logarithms that you use one (usually *log*) or the other and do not mix them. Further discussion of logarithms is to be found in chapter 1 of

Reeds Marine Engineering and Technology Series, Volume 1: *Mathematics for Marine Engineers* (ISBN 978–1-4081–7556–9).

9.4 Useful trigonometry

As we have seen already, the fundamental component of all trigonometric functions is the sinusoidal function as shown earlier in figure 1.2.

The equation: $y(t) = A\sin 2\pi ft$ where $y(t)$ is the vertical displacement in metres (m)

A = amplitude in metres (m)

f = frequency in hertz (Hz)

t = time in seconds (s) and $2\pi ft$ is an angle in *radians*, which when plotted as a graph of $y(t)$ against 't' produces a sine waveform. When the angle is expressed in degrees, the bracket becomes (360 ft).

9.4.1 Common values of sin θ and cos θ

Sin θ and cos θ can have values ranging anywhere between +1 and −1.

sin 0°	= 0	cos 0°	= 1
sin 30°	= 0.5	cos 30°	= 0.866
sin 90°	= 1	cos 90°	= 0
sin 180°	= 0	cos 180°	= −1
sin 270°	= −1	cos 270°	= 0
sin 360°	= 0	cos 360°	= 1

The following formulae are extremely useful in many engineering and physics applications and are used extensively in the Reeds Marine Engineering and Technology Series volumes *Basic Electrotechnology for Marine Engineers* and *Advanced Electrotechnology for Marine Engineers*.

cos A cos B = ½cos (A + B) + ½cos (A − B) where A and B are any two angles in degrees or radians.

sin A sin B = ½cos (A − B) − ½cos (A + B) where A and B are any two angles in degrees or radians.

Area of a circle $A = \pi r^2$. where r is the radius of the circle (m)

Surface area of sphere $= 4\,\pi r^2$

Volume of a sphere $= \dfrac{4}{3}\pi\,r^3$

Example 9.1: Find the surface area of sphere of radius r = 0.5 m (2 decimal places).
Area $A = \pi r^2$ so $A = \pi\,(0.5)^2 = 0.79\ m^2$

Example 9.2: If A = 30° and B = 60°, evaluate cosAcosB (2 decimal places).
Now, cosAcosB = ½cos (A + B) + ½cos (A − B) =
½cos (30 + 60) + ½cos (30 − 60) = ½cos (90) + ½cos (−30) =
½cos (90) − ½cos (30) = 0 − ½ cos (30) = 0.87

9.4.2 Resolving vectors

A vector is a quantity that has both *magnitude* and *direction*, as opposed to a scalar, which has a magnitude only, e.g. a speed of 15 knots is a scalar but a speed of 10 knots on a heading of 090 is a vector (this can be referred to as a velocity of 10 knots on green 090, i.e. speed is a scalar quantity and velocity is a vector quantity). Vector quantities are used in Doppler calculations as it is often the component in a chosen direction rather than the full magnitude of a ship's speed that is needed for the calculations. To obtain a component of a vector, you must resolve it in the required direction. Rules for resolving vectors are as follows:

(i) Find the angular difference between the vector to be resolved and the direction you wish to find the component in. Take the example above: the direction of the vector is 090. Suppose we wish to find the component in the direction 080. The angular difference is (090–080), which is 10°.

(ii) Multiply the magnitude of the original vector by the cosine of the angle between its direction and the required component direction, e.g. the magnitude of the component in the direction 080 is therefore 10 cos10° = 9.85 knots (2 decimal places).

Problems can be clearly represented using diagrams and this is generally the best approach in more complex problems, ensuring the correct calculation of the component's magnitude. For example, sketch the vector sum of two vectors a and b having both direction and magnitude as illustrated in figure 9.3 below.

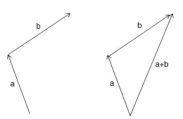

Figure 9.3: *Addition of two vectors a and b.*

9.5 Basic calculator operations

You should ensure that:

(a) you have a 'scientific' calculator, i.e. one that has keys or buttons for functions such as: log, sin, cos, x^y, $x^{1/y}$, etc.;

(b) that you have the correct instruction book for your calculator as there are many differences between various models and makes;

(c) that you have set your calculator to the general default settings of 'scientific' and 'degrees' modes.

The following convention is used. Keystrokes or calculator inputs are shown inside square brackets [ ] and displayed values are shown inside curly brackets { }. Individual digit keystrokes are not shown, i.e. 5.43 is shown as [5.43] and not [5][.][4] [3], where the digits form part of a single number. Results are shown to a sensible, rounded number of significant figures with … representing the rest of the display, which is not given.

9.5.1 Accuracy

Data used throughout most engineering and physics courses will be given to a number of significant figures representative of the accuracy of the measurements taken. The number of significant figures is the number of digits excluding any leading zeros, e.g. 1.23 is 3 significant figures; 456.78 is 5 significant figures; 0.1234 is 4 significant figures; 23.400 is 5 significant figures (trailing zeros are included). When this data is used in calculations, it is important that it is understood that the answer cannot be more accurate than the accuracy of the supplied data. There are sophisticated methods of calculating errors but they will not be considered here. Only a simple method of assessing the accuracy of a result will be used here, as follows: Examine all the data used in the calculation and state the final answer to no more significant digits than the least accurate data given. Do not round intermediate answers as this can lead to large subsequent errors where several successive calculations are used to arrive at the final answer.

9.5.2 Adding, subtracting, multiplying and dividing

Make sure you can carry out the previously covered basic operations correctly. When the calculation involves several terms using multiplication and division, you must remember that the order of the terms on the top and the order of the terms on the bottom does not matter, but it is very important to divide all the terms on the bottom into all those on the top. The process involves computing the products on the top and then dividing all the terms on the bottom into that product. This is where brackets are useful.

E.g. $27 \times 1.6 \times 3.2 \div 7.9 \times 0.51$ can be written as

$((27 \times 1.6 \times 3.2) \div (7.9 \times 0.51)) = 34.31$ (2 decimal places)

The sequence of calculator operations is therefore exactly as written, i.e.

[(][27][×][1.6][×][3.2][)][÷][(][7.9][×][0.51][)][=]

which results in your calculator displaying the answer {34.31}.

You can carry out the calculation without using brackets but you must remember to replace all [÷] terms on the bottom with [÷], i.e. the calculator sequence is:

[27][×][1.6][×][3.2][÷][7.9][÷][0.51][=].

It is not the result of the sequence [27][×][1.6][×][3.2][×][7.9][×][0.51][=]. A highlighted $\times$ means that the result of all the previous operations is multiplied by 0.51, not divided by 0.51, which is what the initial expression means. Try it yourself and you will see that it gives a different answer from 34.31.

9.5.3 Use of standard form

As has been shown previously, very large and very small numbers are best expressed using powers of ten. Powers of ten are entered using the EXP key (EXP is short for exponent), e.g. 6.62×10^{-34} (the value of Planck's constant) is entered as [6.62][EXP][34][±] and displays as {6.62⁻³⁴}. You will need to enter the minus sign for the exponent using the ± key and note that the calculator does not display the × 10.

Suppose we have to evaluate: $6.62 \times 10^{-34} \times 3.00 \times 10^{8}$ divided by: 0.550×10^{-9}

The correct calculator sequence will be:

[6.62][EXP][34][±][×][3.00][EXP][8][÷][0.550][EXP][9][±][=]

The display sequence will be as follows:

{6.62}

{6.62 ^{00}}

{6.62 ^{34}}

{6.62 $^{-34}$}.................. at this point you have entered (6.62×10^{-34}) into the calculator

{6.62 $^{-34}$}

{3.00}

{3.00 ^{00}}

{3.00 ^{08}}

{1.986.... $^{-25}$}............. on pressing ÷ you have instructed the calculator to carry out the previously entered multiplication of (6.62×10^{-34})

by (3.00×10^{8}) and the result is then displayed.

{0.550}

{0.550 ^{00}}

{0.550 ^{09}}

{0.550 $^{-09}$}

{3.61…$^{-16}$}…on pressing [=] you have instructed the calculator to divide by (0.550 × 10^{-9}) and the result is then displayed. This final result must be written down as 3.61…. × 10^{-16} and not as 3.61…$^{-16}$ as this means (3.61 …)$^{-16}$ – which is, of course, a quite different number.

9.5.4 Squares and square roots

(i) A number multiplied by itself is said to be *squared*, e.g. $2 \times 2 = 2^2$ (2 squared)

Any number x multiplied by itself is therefore x^2, and your calculator will have an x^2 key on it. On some calculators the x^2 and $\sqrt{x}$ functions share the same key and one of them will have to be accessed by using the [INV] or [2nd F] key – see your calculator manual.

For example, to find the value of 5.9^2, the following keystrokes are needed:

[5.9][x^2] or [5.9][INV][$\sqrt{x}$], depending on your calculator. Note that you do not need to press the [=] key to obtain the square ($5.9^2 = 34.81$).

(ii) Suppose you are given the number 34.81 and are asked the question 'What number multiplied by itself gives the answer 34.81?', i.e. you are being asked to find the square root ($\sqrt{}$) of 34.81. This is done by using the [$\sqrt{}$] key.

To find the value of $\sqrt{34.81}$, the following keystrokes are needed:

[34.81][$\sqrt{x}$] or [34.81][INV][x^2], depending on your calculator. Again, note that the [=] key is not needed.

9.5.5 Use of the x^y key

This key enables you to raise any number (x) to any power (y). On some calculators you will find that the x^y and $x^{1/y}$ functions share the same key and one of them will have to be accessed by using the [INV] or [2nd F] key, as for x^2 and $\sqrt{x}$ functions.

You could use this key to find the value of 6.7^2, for example;

[6.7][x^y][2][=] or [6.7][INV][$x^{1/y}$][2][=] gives the value of 6.7^2.

The main use is for finding higher powers such as cubes, 4th powers, etc.

4.56^4 is evaluated by [4.56][x^y][4][=]   (result: 432... ...)

$3.45^{2.7}$ is evaluated by [3.45][x^y][2.7][=]   (result: 28........)

$6.78^{-1.6}$ is evaluated by [6.78][x^y][1.6][±][=]   (result: 4.67 x 10^{-2})

9.5.6 Use of the $x^{1/y}$ key

This key enables you to find the (yth) root of any number (x). The (yth) root of (x) can be written in two ways: $^y\sqrt{x}$ or $x^{1/y}$. The examples that follow show how the $x^{1/y}$ key is used. To evaluate the $^2\sqrt{21}$, enter:

[21][$x^{1/y}$][2][=] or [21][INV][x^y][2][=] (as this can be written as $21^{1/2}$).

This second method gives the same result as using the $\sqrt{x}$] key.

Now try the following examples:

(i) $21^{1/5}$ (answer: 1.83842)

(ii) $(34 \times 10^3)^{1/3}$ (answer: 32.3961)

(iii) $(2.34 \times 10^{-3})^{1/2.1}$ (answer: 0.558. ...)

9.5.7 Logarithms and antilogarithms

As has been shown before, any number (y) can be expressed in the form (10^x), i.e. $y = 10^x$.

Remember, (x) is called *the logarithm to the base 10 of (y)* and (y) is called *the antilogarithm of (x)*.

On most calculators you will find that the log and 10^x functions share the same key and one of them will have to be accessed using the [INV] or [2ndF] key, as for the previous functions discussed.

(a) To find the logarithm of 123.45, for example, the following keystrokes
are used:

[123.45][log] or [123.45][INV][10ˣ] giving 2.09…

To evaluate the logarithm of ([27.3][÷][6.7]), compute the bracket first and then the logarithm, e.g.: [27.3][÷][6.7][=][log]

Note that the [log] key acts directly on the displayed value and the [=] key is not needed.

This is where brackets are useful. The key sequence above gives the same result as the following key sequence [([27.3][÷][6.7])][log] (answer: 0.610 …). The second sequence allows you to match your keystrokes more closely to the expression to be evaluated. Here, the closing of the bracket)] carries out the [=] function on the bracket's contents. Brackets will also prevent you from entering the incorrect key sequence [27.3][÷][6.7][log][=] which gives 33.04…, the [log] acting on the displayed value of {6.7…} and not the result of (27.3 ÷ 6.7).

Note that the logarithm of a number between 0 and 1 is negative, i.e. logarithms of fractions are negative (remember the negative sign). The logarithms of 0 and negative numbers will yield the 'ERROR' message on the calculator.

Now try:

(i) log (6.41 × 3.72)   answer: 1.38 (2 decimal places)

(ii) log (1.4 × 3.7 × 10^4)   answer: 4.7 (1 decimal place)

(b) To answer the question: 'What is the number whose logarithm is …', we must
find the antilogarithm using the [10ˣ] key. For example, suppose we know that
the logarithm of a certain number, y, is 3.2, i.e. log y = 3.2. To find y we need to
carry out the key sequence:

[3.2][10ˣ] or [3.2][INV][log] to get the result 1584.8932 displayed.

Now try:

(i) 0.260 = log p   p = 1.82 (2 decimal places)

(ii) 2.57 = log p   p = 371.54 (2 decimal places)

9.5.8 Sine, cosine and tangent

The [sin], [cos] and [tan] keys are used to find the sine, cosine and tangent of any angle respectively. The angle may be expressed in *degrees* or *radians* so you must ensure that your calculator is set to operate in the correct mode before you begin. Usually the angle required will be in degrees, and the default expression will be in degrees also.

If a [Mode] key is available, use it to select (Deg) operation. Some calculators have a key, or keys, labelled with Deg and Rad. Other functions like Grad or Stats may also be present on some calculators. Some may also show in the display which mode they are in after you press one of the mode change keys.

(a) To find the sine of 56.7°, the following key sequence is used: [56.7][sin] giving {0.835…} Note that the [=] key is generally not needed.

For angles greater than 180° the result will be negative, e.g. the sine of 199° is −0.325… The negative sign is important and must not be ignored. This result is given in fractions of a degree, not minutes and seconds. Some calculators can convert between these units.

The [cos] and [tan] keys can be used similarly.

(b) To find the angle given the sine of that angle requires the use of the inverse sine or sin^{-1} key. This is normally a second function of the [sin] key and is accessed by using the [INV] key, e.g. sin θ = 0.5. To find the inverse sine, we use the key sequence [0.5][INV][sin], giving {30°}.

Similarly if sin θ = − 0.2419219, we can find θ with the key sequence:

[0.2419219][±][INV]

Further discussion of trigonometry is to be found in chapter 7 of Reeds Marine Engineering and Technology Series, Volume 1: *Mathematics for Marine Engineers* (ISBN 978–1–4081–7556–9).

9.6 Self-assessment questions

After studying this chapter, you should be able to answer the following questions.

1.   Expand $(a + b)^3$

2.   Expand $(a + b)^2 \times (c + d)$

3.   What is the value of $a^n \times a^m$ where $a = 3$, $n = 4$ and $m = 2$ (4 decimal places)?

4.   What is the value of $a^n \times a^m$ where $a = 2$, $n = 3$ and $m = 4$?

5.   What is the exact value of $a^n \times a^m$ where $a = 1.6$, $n = 2.4$ and $m = 3.3$ (2 decimal places)?

6.   Demonstrate the scientific calculator functions discussed in this chapter.

7.   Rearrange the equation $y = mx + c$, making x the subject of the formula.

8.   With the use of your scientific calculator, what are the values of the following:

(i) $1.9 \times 2.5 \times 2.5 \div 5.9 \times 0.63$ (5 decimal places)
(ii) the logarithm of (4.21×5.62) (5 decimal places)
(iii) $3.61^{5.6}$ (2 decimal places)

9.   Find the gradient (dy/dx) for $y = (a + x)^2$

10.   Find the derivative of $f(x) = \ln(3x - 4)$

10

Student Centred Study Skills for Further and Higher Education

'Success is the ability to go from one failure to another with no loss of enthusiasm.' Sir Winston Churchill

10.1 Preparing for exams and general study support

My first brush with the 'wrong end' of Teaching and Learning was to hear my parents informed, when I was aged 6, that I would never learn to read! However, having learned to read through the perseverance of my mother and my own hard work, I subsequently went on to write several books and through my work meet many interesting people, including members of the peerage and famous scientists (including a Nobel Prize winner). This just goes to show that people who make authoritative predictions on your future can get it completely wrong!

Absolute statements of the above type exclude any capacity for change and improvement, and are clearly a very foolish thing for a teacher to declare. A teacher is a huge factor in both building up and tearing down a student's self-confidence and should consider themselves to be a mentor, capable of drawing out from the student latent skills and abilities that they probably could not see in themselves. However, it is clear that the student must be at the heart of their own learning experience and not just an outside observer watching what happens to them.

It was for these reasons that I (CRL) felt compelled to include a chapter on Study Skills, as this is really important for those studying at Further Education (FE) and Higher Education (HE) levels, and it is even a valuable area for those not studying for further qualifications. It may prove to be the most important chapter to the reader for subsequent study. It is assumed too readily that students already know 'what to do' when it comes to study, irrespective of their particular educational/training background. This assumption leads many students, particularly overseas students, to spend a considerable amount of unnecessary time 'in the dark' as to what type of problem/task they have been set, what they need to do, how to go about solving the problem and how to develop a plan of action and the consequent 'tools' and 'skills' they are likely to need to achieve this aim. It is better for students to have some sort of generic plan or process in place that can be adapted as necessary to solving different sorts of problems as they arise. It is in this context that the help of a tutor can prove valuable and as such there is a section discussing the role of a tutor.

In this chapter we will look at general principles for scientific essay work, and the sort of things that will lead to an overall good rather than bad assessment being made, providing a balanced viewpoint from both a teaching and undergraduate student's perspective.

We provide a checklist appropriate to producing not only essays but also information leaflets, PowerPoint presentations, etc. We will also look at some ways a student may best prepare for this. For those wishing to explore a more rigorous pedagogical approach, there are many books and web-based resources currently available [10.1–10.2].

10.2 General principles for written scientific work

To write a good essay or report, you need to have a clear structure. To do this, you must follow a clear logical development of concepts as your analysis of the topic proceeds and you should include the following:

10.2.1 The title

Make sure you really understand what it is asking you to do. If you have the freedom to choose the title of your essay, make sure it is clear and unambiguous. If you are unclear about what you are being asked to do, then go back and ask. A little time spent clarifying what you are required to do can prevent you from going off course and save an enormous amount of effort in the long run.

10.2.2 Background reading

Before you start writing anything, gather together your relevant sources. Read the actual references suggested and start putting forward ideas of your own to others, which can be weighed or tested before you start committing anything to paper as submitted work.

10.2.3 Introduction

The use of an introduction puts the topic in context and 'signposts' clearly where the essay, information leaflet or study topic is going. A good essay will flag up in the introduction where it is going, and may even state any extremely important or valuable findings.

You should have a clear idea in your own mind of where you are going with the essay and how the various components will fit with the central strand of your discussion, in the same way as you would expect to have a passage plan prior to your departure from port. Always provide evidence that supports your statements, be that evidence from the literature or from your own personal experience.

10.2.4 Main text

Try to be clear and concise in your words, and try to retain the view of an impartial observer. Be objective.

Use illustrations and diagrams that help explain points of text or provide 'added value'. Including pretty pictures that do not really help or add anything should be avoided at all costs. If you use diagrams, make sure you label them and refer to them appropriately in the text. Make sure all figure text is legible!

Offer new insights, evaluate ideas and findings, explain any observed phenomena. Show what you can do with the information you have obtained.

10.2.5 Subheadings

Use subject headings to distinguish between different parts of the essay or report. Make sure your text does not become separated ('widowed') over a page from its title or subheading.

10.2.6 General mistakes

Common essay mistakes involve the use of massive blocks of text. Try breaking up large sections of text into smaller ones. Avoid using abbreviations that are not defined. Do not assume the reader knows the topic you are talking about unless you have been told to make particular assumptions about the target audience. Failure to check for spelling mistakes and grammatical errors will result in a poor overall mark for your assessment. This is where the use of spell-checking software – and a tutor – can be extremely useful. The reader should set the language to their specification, because the spell-checker will recognise an 'incorrect' English (UK) spelling if the language is set to English (US) and 'correct' it. Even when this is done, some frequently used words, such as discrete/discreet or practice/practise, may be used in the wrong context and will not be spotted by the spell-checker. It is thus advised that the reader does not rely solely on the benefits of spell-checking software!

10.2.7 Conclusions

This section provides a brief summary of *your* main points and any general conclusions you have found and drawn yourself from the material. This is important, as the introduction and the conclusion will be the two parts of the essay that your tutor and examiner will remember the most. It is a great opportunity to provide your strongest evidence and to include your own reasoned ideas.

10.2.8 References

It is important for you to show the references for your work. Plagiarism, the submission of copied work or shared work (not joint project work), is unacceptable. If you quote from books, journals, the Internet or other sources, make sure you have suitably referenced the material using the standard referencing method that has been requested.

10.2.9 Checking the essay and post-submission issues

Check the final version by printing it out and reading it through aloud. If writing the file using electronic media, make sure you save the file at regular intervals in two or more locations and on more than one type of media, if possible.

Sometimes a formative assessment may be requested by your tutor. This is an assessment that will be examined critically to provide you with unrecorded feedback. Formative feedback is entirely for your own benefit and you should take advantage of this wherever possible. Summative feedback is submitted for coursework marking. If you have work that only has a final summative deadline date you should still consider asking your tutor for some formative feedback so you can make any revisions prior to the summative submission of the work.

After submitting the document, make sure you have both a hard copy and a saved 'soft copy' e-version yourself – it may be needed again at a later date. With any project, it is really useful to keep a specific folder that holds all the essential aspects of this project work. What is the minimum material you need to reconstruct the original work? This should be kept in your folder until you are certain you will never need it again! Then 'weed', delete or throw away unnecessary materials. This is an area I (CRL) have particular trouble with; it is not sufficient to think the job is done until all the 'waste' has been cleared away and things put back where they belong. Give a copy of your essay to as many Faculty members as possible who could make a positive contribution, as this will help you to find out aspects of the project that you hadn't considered at first. *Plans will generally fail for lack of advice.* The more *good* advice you get, the better your work is likely to be.

10.3 Critical points to address to achieve the highest essay marks

One checklist I (CRL) have developed over the years has helped students achieve the higher marks by being self-critical of meeting the required standards. This, I hope, will prove useful to the reader as well.

10.3.1 Introduction

This section requires a clear and well-defined title. The essay/information leaflet/ PowerPoint presentation should be in the form of a brief general statement that leads the reader into the chosen topic. Most projects have a clear aim in mind, details of which are usually provided to students within the course or module record. Module records will contain useful information such as: aims, learning outcomes, assessed skills elements, etc. For example, a learning outcome might be to evaluate the successes and weaknesses of your work. An assessed skill element might be developing ability at report writing, or an opportunity to demonstrate creativity and original thought. The introduction shows how the specific topic is significant in the subject area of interest (applicable much more widely than just in physics and engineering subjects!).

10.3.2 The essay

We will now look at the contrast between a good and a bad essay. A real essay will, of course, be likely to exhibit some of the characteristics of both categories, but discussed together should provide the student with some degree of understanding of what is a good essay and the work that this entails.

10.3.3 The good essay

A good essay or other such written task will exhibit the following attributes:

Presentation and overall layout

The essay or leaflet is eye-catching and immediately communicates that it is descriptive and to the point. The layout is uncluttered with a pleasing appearance.

There is strong evidence of creative, but not excessive, use of colour, fonts, graphics and text. There are no grammar, spelling or punctuation errors. The information chosen has been produced using the correct format and has the correct number of pages that are required.

Information content

The essay/leaflet should typically be at a technical level understandable by peers. The information content is factually accurate. There is a logical and consistent progression of the information presented. The essay identifies and explains relevant key principles, e.g. 'the principles of the maritime sensor technology' are discussed. The essay makes good reference to the maritime technology used in relevant context. The information content is presented in an efficient manner using tables, boxes or other appropriate means.

Range of materials and explanatory qualities

The range of material presented is highly relevant to the topic, well explained and is well selected and organised. The material presented is clearly explained, with many technical terms clearly defined. The student material shows strong evidence of intelligent comment on the material presented. For example, the operational role of the maritime sensor technology is also clearly explained.

Conclusions

The essay is summarised and concluded effectively. The aim of the essay is maintained throughout and clearly achieved at the end. A good essay also demonstrates that the project has drawn out your abilities to conduct scholarship or research. A project or task should always be viewed as another opportunity to develop communications skills with staff and other students.

10.3.4 The bad and the ugly essay

A poor essay/information leaflet/PowerPoint presentation will generally have the following characteristics in common and should be avoided at all costs!

Introduction

A poor essay will usually have an ambiguous, imprecise or long-winded title and will not have a logical introduction to the chosen topic. Its range, scope and aims will be extremely unclear. The introduction will be poorly defined and the overall aims of the work will be vague. Poorly organised or 'muddled' thinking at the introduction stage will influence the outworking of the rest of the essay.

Presentation and overall layout

The submitted essay or student information leaflet will have an uninteresting appearance and it will be difficult to see initially (or at all) what the content is about. The essay is cluttered with information or is overly crowded. The leaflet displays little evidence of creativity and if required in colour will have been submitted in black and white. The leaflet has a large number of grammar, spelling and punctuation errors. The wrong format has been used and the essay does not contain the required number of pages.

Information content

The leaflet or presentation is too simplistic or too advanced in content. The student has not considered carefully enough the level of his target audience. Material will be at the wrong level and poorly explained. The information content is inaccurate and contains many factual errors. The leaflet has no logical or consistent structure and no gradual progression of concepts or ideas. The leaflet makes little or no reference to key principles and concepts involved in the engineering and physics technology. The essay makes little or no reference to the maritime sensor technology used in a relevant context. The information is presented in an inefficient manner, wasting much of the space available. A poor collaborative piece of work will have little or no evidence that the material was a team effort, with no explanatory contributions from the various team members.

Range of materials, technical accuracy and explanatory qualities

There is a very narrow range of material presented with much being irrelevant, poorly selected or disorganised. The material is poorly explained with many technical terms left undefined. The leaflet presents the material entirely in a repetitive fashion, without intelligent comment. There is no explanation of the sensor technology used in its operational role. A collaborative piece of work, where required, will exhibit clear evidence that the explanatory qualities were the work of a single team member with little or no contribution from other members.

Conclusions

There is no obvious summary, or it is either too short or fails to establish an effective conclusion. The leaflet fails to achieve a clear aim.

10.4 Engineering and physics study tips for studying and exams

10.4.1 Modes of learning

Most educational and training programmes will expect you to be or to become an independent learner as quickly as possible. Thus you will need to think extremely carefully about your own learning style and how best you can use your learning strengths while completing your academic studies. It is often reported that learners arriving at Further Education (FE) colleges have insufficient skills in maths and science. Some learners arrive with some or all of the maths and science prerequisities but are still unable to cope well with the technical requirements of their chosen vocational or academic study programme.

There are different sorts of learners and therefore there are different sorts of learning styles that best match the individual. In addition, every learner will have different needs and aspirations, which will need to be met if the student is to gain the most from their course of study. However, real people often have attributes from several of these different styles – as they are, in fact, unique and do not really fit into a single category!

However, we can, for the purposes of argument, break down a learner into one of three learning types: auditory, visual and kinaesthetic.

An *audio* learner is one who learns best by hearing information, often through verbal instructions from others.

A *visual* learner is one who learns best from seeing information, presented in text, diagrams, pictures or charts, typically by watching demonstrations.

A *kinaesthetic* learner will learn best by the 'doing' of tasks, by making, building and moving things. Kinesthetic learners benefit most from practical involvement in a task.

How a learner achieves these ends will typically be through some of the following modes: spelling, reading, handwriting, memory, imagery, problem solving, emotions and communication. For example, consider imagery: a visual learner will have a vivid imagination, thinks easily in pictures and can visualise even very fine detail in a topic. An auditory learner will subvocalise and tends to imagine things in terms of sounds, while the detail found by the visual learner will be less important to them. To a kinaesthetic learner, imagery is generally unimportant unless the imagery is relevant to the actualising or movement that is required.

Depending on the individual learner, their level of subject engagement should be obvious. Is the student distracted? How does the student respond to long periods of activity/inactivity? How are they responding to new situations?

A real learner, as I have said, is often a combination of all three of these learner types – meaning that, whether we realise it or not, we are in fact participating in multisensory learning.

It has been said that on average a person will remember:
- 10% of what they read
- 30% of what they see
- 50% of what they see and hear
- 70% of what they discuss with others
- 80% of what they experience personally, but
- 95% of what they teach others

Which means that, in a multisensory environment, this amounts to most of what you read, hear, see, say and do! Much more detail on this subject is to be found in pedagogical literature already mentioned [10.1–10.2].

10.5 Learning approaches

Here are some 15 ideas that may work well for you in terms of approaching your learning.

10.5.1 Planning your day

When do you get the best results, the best understanding; when do you achieve your best work productivity? This should be the time we safeguard from other distractions. Some people work best early in the morning, others late in the evening, or – like James Clerk Maxwell – well into the early hours of the morning (without disturbing other people who may be asleep, of course!). If you are always late for class, you should consider improving your time management. Don't leave things to the last minute. Don't put off coursework or assignments to a later date. It is very hard to finish a piece of work that you do not start. Create a study/revision plan well in advance of when you need it.

10.5.2 Spaced repetition

This learning technique involves breaking up information into relatively small chunks of material and reviewing them consistently over a long period of time – not just in a panic an hour before the exam.

10.5.3 Tell a story

Turning the details you need to remember into a story can help make the information more meaningful and may involve inventing a mnemonic. A mnemonic is a tool used to help remember facts or a large amount of information. It can be in the form of a song, an acronym, a rhyme or a key phrase that helps you to remember a list of facts in a certain order. For example, if you want to remember the colours of the visible spectrum ROYGBIV (Richard Of York Gave Battle In Vain), or factors of loss SAS (Scattering, Absorption, Spreading).

10.5.4 Location

Some research suggests that studying the same things in a different physical location on different days can make students less likely to forget the information they are trying to learn. This may be because you are now forcing your brain to form new mental associations with the same material, so that it becomes a stronger memory. Where you study is also important for several reasons. Is it a comfortable place: is it too hot or cold, is the seating comfortable, how good is the lighting, is it free from distractions? Does the location help the process of studying, making it easier? Sitting in front of a widescreen TV in the main university student social area is probably not going to help your learning as much as other possible locations. If possible, block a specific time and location for your study, and before you start to study have as clear an idea as possible of what you are trying to achieve in this time. If possible, having an objective that you can measure your output or productivity against is also worthwhile – for example, 'I will work through three maths calculation sheets today'. If at sea, plan to work in your cabin when you are least likely to be distracted by other cabin mates.

10.5.5 Multiple maths tasking

Don't always stick to a single topic; try instead to study several different techniques that work together in one sitting (synergism) or do a series of problems that requires you to think and differentiate between them – for example, doing calculations involving divisions, multiplications, integration or differentiation means you must stop and think about which mathematical approach is best to use.

10.5.6 Practising exams and testing yourself

Make sure you practise exam papers and question topics fully before the exam. Create your own questions from source questions you may have been set and examples you have been given (including those in this book); this will help you to reinforce what you have learned already. Quizzing yourself and doing a mock exam are one of the best possible ways to prepare for a real exam. It is harder to remember information in practice exam mode than during conventional study, so the more you practise mock exam conditions, the more likely you will be to remember information in the future if you develop an active exam-focused approach. This is true of more academic subjects or practical chart-plotting exercises.

10.5.7 Write it out!

Research suggests that we store information more securely in the brain when we write it out by hand compared with when we type it or just repeat it verbally. Start by recopying the most important notes of your course on to a new sheet of paper. If you do not know how to take notes in class, seek further advice from your engineering or physics tutor. What about using a dictaphone or digital tape recorder? Tutors should also be able to help you to develop memorisation techniques.

10.5.8 Drawing

Drawing a diagram can prove to be very useful in order to summarise a complicated topic. Memorising a diagram can often help build memory associations in many different topic areas.

10.5.9 Reading aloud

Despite what I have just said, reading information out aloud will help you to store it mentally in at least two ways: by seeing it and hearing it. Remember the multisensory environment!

10.5.10 Group work

Find a few hard-working students and get together every few days to review material and swap strategies for solving problems. This should not replace your commitment to making your own effort to solve problems, as without it you will not develop the skills you need for the future.

10.5.11 Take a break!

Taking breaks will help sustain you for the long haul. Go for a walk, eat something or make yourself a drink (and always drink plenty of water). Regular breaks can boost productivity and improve your ability to focus on a single task. Just stop working at the laptop for five minutes every hour; this will also help to prevent eye fatigue. Some research suggests that coffee, tea and energy drinks may help to keep people alert in the short term due to their caffeine content, but excessive use should not be seen as a replacement for regular exercise, rest and sleep. Similarly, if you really want to make progress then having all your social media, phones, iPad and laptop on the go when you are trying to study is probably a bad idea.

10.5.12 Background music

It is understood that music can help us cope with stress and reduce anxiety and tension. Some studies have shown that classical music helps both children and adults to learn better and remember facts and lists of items better. It is unlikely, however, that music with a fast beat or heavy metal will achieve the same educational ends; everyone has a different taste in music, but we are considering here the likely best educational environment that will support academic study.

10.5.13 You are what you eat

Good nutrition and diet is essential to healthy thinking. In addition, omega-3 fatty acids, found in certain fish, nuts and olive oil, are well known for their brain-boosting potential. One study claimed that eating a balanced combination of omega-3 and omega-6 fatty acids before exams reduced test anxiety.

10.5.14 Timetable

Organise your topics and modules into a clear timetable for progression throughout the course of the term and year, and crucially have a timetable for your final degree or course exams.

10.5.15 Review your learning before you go to sleep

Apparently the brain strengthens new memories during sleep, and especially the last thing you have remembered at the end of the day. However, there is little point trying to do this when you are exhausted from overwork, so when you are starting to get tired, listen to your body – it is probably a good guide for when to stop!

10.5.16 What actually works?

After trialling all the ideas discussed in this section, you should find some that work better for you than others. These are clearly worth continuing, while those that do not are probably best set aside for now – although your learning preferences may change over time. You may also find that you have a tendency to put off the hardest things, which you know are difficult. Unfortunately, such tasks will never get any easier unless you address this problem. This may mean seeking additional help and support, perhaps first with groups of other students or with staff, as necessary.

The role of an educational mentor or tutor

It is crucial for the student to find a good tutor or mentor. If no one is able to fulfil that role (perhaps you are studying while at sea), see if you can work with and learn from more senior colleagues. A good tutor should provide sufficient numeracy and English support to each learner, with patience, especially if the student has already struggled with a numeracy-related lack of skills in the past. The tutor and student should understand the value of an individualised learning programme, to identify both weakness and strengths in the skills of the learner. They can then work at overcoming any problem areas while building upon the learner's existing strengths. If possible, the learner should take an integrated approach, with the support of the tutor, to incorporate the same mathematical and numeracy skills into several subject areas. For example, doing mental arithmetic in the course of daily activities such as shopping, rather than just waiting for the bill at the checkout, is a good first step – but not always applicable at sea!

It may be necessary to get to the root of common misconceptions in mathematics and in scientific understanding. It is critical for the student to understand where they have gone wrong and why and to apply the correct reasoning in future. Hence, the approach taken throughout this book and others in the Reeds Marine Engineering series and Introductions series is to provide extensive worked examples and further study questions to practise.

Students should experiment themselves with different types of activity such as working in pairs or groups, rather than just trying to work on their own. This will allow students to explore mathematical, engineering and physics based problems and calculations to develop their understanding and confidence. Students should use as wide a range of resources as possible, making use of Information Technology (IT), the internet and software based packages to develop their ability to manage their own learning independently and in a way that suits them. Problem solving should be seen as an opportunity to apply their growing numeracy skills and confidence to attempt increasingly harder real world problems.

The student, with the tutor's assistance, should try to assess their abilities at the start of their programme to provide them with initial feedback on their current levels of ability and numeracy in relation to their specific learning programme, career aims and future employment. This is critical as the student should be nurtured from the start, and guided as clearly as possible. It should not be the case that a student discovers their particular weaknesses after the end of year examination!

REFERENCES

[10.1] Pedagogic Research Institute and Observatory at the University of Plymouth. PedRIO http://www1.plymouth.ac.uk/research/pedrio/Pages/default.aspx

[10.2] *Learning and Teaching in Higher Education The Reflective Professional,* Greg Light and Roy Cox (Paul Chapman Publishing, London, 2009, ISBN 978–1-8486–0008–9)

For Teaching and Learning questions associated with Marine Engineering and Physics, Dr Chris Lavers may be contacted at: christopher.lavers@plymouth.ac.uk. I am particularly interested in individual case studies or vocational examples and any solutions you may have found to address maritime problems you have encountered.

Appendix A

Useful data

This section introduces commonly used wave data, multipliers and Greek letters.

360°	$\equiv 2\pi$ radians
1 radian	$\equiv 57.296°$

Speed of EM waves in free space	approximately: 3×10^8 ms^{-1}
Speed of sound in air	approximately: 330 ms^{-1}
Speed of sound in water	approximately: 1500 ms^{-1}

Standard multipliers:

Giga (G)	$= \times 10^9$
Mega (M)	$= \times 10^6$
kilo (k)	$= \times 10^3$
milli (m)	$= \times 10^{-3}$
micro (µ)	$= \times 10^{-6}$
nano (n)	$= \times 10^{-9}$
pico (p)	$= \times 10^{-12}$
femto (f)	$= \times 10^{-15}$

Some commonly used Greek letters:

Alpha	α
Beta	β
Gamma	γ
Delta	δ, Δ
Epsilon	ε
Lambda	λ
Phi	φ
Sigma	σ
Theta	θ
Tau	τ
Omega	ω, Ω

Appendix B
Properties of Water

Water is one of the commonest substances to be found on Earth and has several unique physical and chemical characteristics.

B.1 Ocean surface salinity

Salinity affects density and therefore the buoyancy of ships and submarines. Seawater consists of a small variety of salts dissolved in water, the global average being about 3.5 per cent or 35 grams per kilogram, or parts per thousand by weight. Salinity is defined in terms of a ratio and as such does not have a unit. Salinity values around the world's oceans vary from 35, due to either the removal or addition of fresh water. Away from coastal waters, the balance between evaporation and precipitation controls overall salinity.

Salinity is raised if fresh water is removed by either high evaporation or freezing of ice. Salinity is lowered if fresh water is added either by precipitation, river outflow or melting ice.

Enclosed seas can exaggerate the effects of these mechanisms. For example, in the Dead Sea salinity can rise to as much as 210 per thousand due to evaporation, while salinity in the northern Baltic can fall to as low as 2 parts per thousand under the influence of spring meltwater.

Typical ocean and sea salinity values are given in table B.1.

Sea or ocean	Salinity in parts per thousand
Baltic	2–10
Black Sea	18
Equator	36
Subtropics	37
Western Mediterranean Sea	38
Eastern Mediterranean Sea	40
Red Sea	40
Dead Sea	210
Global average	35

Table B.1: *Typical ocean and sea salinity.*

The six most abundant (by weight) of ions in sea salts are also given in table B.2.

The most abundant constituent of seawater is the negative anion (chloride) Cl⁻, the ionic form of chlorine; the second most abundant is the positive cation sodium Na⁺.

Ion	Gram weight per kg	% by weight
Chloride	19.111	55.04
Sodium	10.629	30.61
Sulphate	2.667	7.68
Magnesium	1.281	3.96
Calcium	0.403	1.16
Potassium	0.383	1.10

Table B.2: *Typical ions found in water with their grams per kg.*

B.2 Ocean surface temperature, pressure and movement

Ocean surface temperature rarely exceeds 30° C, although in shallow coastal waters can rise to as high as 42° C. In the Arabian Gulf, summer sea temperatures typically reach 34° C. Global minimum temperatures are found around the Antarctic, where they can drop to as low as –1.3° C. Sea Surface Temperature (SST) isotherms follow general latitudinal patterns, although distortions can occur on the continental margins where the confluence of major warm and cold currents disturb this pattern. Poleward of latitude 60 degrees, there is little seasonal change in temperature due to the low angles of incident solar radiation (insolation). Daily sea surface temperature changes are generally small, except in shallow water.

Pressure There is an overall increase in pressure of 1 atmosphere or 1 bar pressure for every 10 metres descent into the seawater. This 10 m depth is equivalent to the entire pressure increase from the water surface to the edge of space.

Movement of salt water. There is horizontal movement, caused mainly by tides and the wind, and vertical movement due to various convection currents. Tides are caused by gravitational attraction to other bodies in space – the main body being the moon.

B.3 Chemical Characteristics

pH Seawater has a remarkably constant pH (or hydrogen valency) of 8, i.e. it is slightly alkaline and acts as a 'buffer' against large pH changes. Freshwater, however, is a poor buffer solution and varies with the amount of dissolved carbon dioxide pH life range 4.6–8.5, with pH = 7 being neutral with equal numbers of positive and negative ions present in solution.

Dissolved gases Seawater is saturated with oxygen (34%); other gases present are nitrogen (65.6%) and carbon dioxide (0.3%). Note the higher oxygen content of seawater than in normal air and the approximately equally quantity of carbon dioxide in the world's oceans. The oceans provide a significant *carbon sink*, soaking up carbon dioxide produced by both natural and anthropogenic means, and consequently any change to the ocean's ability to continue to absorb carbon dioxide from the air will have a large impact upon climate change modelling predictions. Further details on the properties of water can be found in Appendix 31, 'Properties of Water', in Reeds Marine Engineering and Technology Series, Volume 13: *Ship Stability, Powering and Resistance* (ISBN: 978–1-4081–7613–9).

Appendix C

Differentiation

Differentiation is the process of finding the gradient of a curve. The gradient or slope (derivative) of an equation (or function), e.g. $y = 2x$, is the rate of change of the equation determined by a changing quantity (or variable) x.

It is a fundamental tool of mathematical *calculus* developed by Sir Isaac Newton (1643–1727) in his quest to understand the force of gravity. In dynamics, the gradient or derivative of the position of a moving object with respect to time yields the object's velocity, a measure of how quickly the object's position changes over time. The gradient, however, measures the instantaneous rate of change of the position, and not the average rate of change. The process of finding the derivative is called *differentiation*. The reverse process is most commonly referred to as *integration*.

C.1 How to find the gradient or derivative of a curve at a given point

Take a point Q on a curve whose x co-ordinate is just slightly greater than P for the equation $y = x^2$ (see figure C.1).

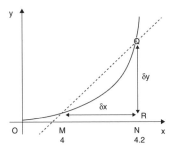

Figure C.1: *Finding the gradient of a curve at a given point.*

Draw the line QR, with PR drawn parallel to the *x* - axis.

Now take the distance PR (the difference in the *x* co-ordinates) as 0.2

Therefore, if the x co-ordinate of P is 4, the x co-ordinate of Q (i.e. ON) = 4.2

As Q lies on the curve $y = x^2$, the y co-ordinate of Q = 4.2 × 4.2 = 17.64

Hence QR = QN − RN = QN − PM = 17.64 − 16 = 1.64

Hence the gradient of the chord PQ = QR/PR = $\dfrac{1.64}{0.2}$ = 8.2

Now let Q approach P by taking smaller and smaller values for PR. In each case, obtain the gradient of the chord PQ as above.

For simplicity, results are given in table C.1.

PR	0.2	.02	.002	.0002
ON $x + dx$	4.2	4.02	4.002	4.0002
QN x^2	17.640	16.1604	16.016004	16.0016
QR	1.64	0.1604	0.016004	0.0016
QR/PR	$\dfrac{1.64}{0.2} = 8.2$	8.02	8.002	8

Table C.1: *Increasing accuracy calculation of gradient.*

It is clear from Table C.1 that as PR decreases and approaches the value zero, the ratio QR/PR approaches ever more closely the value 8.

Hence the gradient of the curve at P = $\displaystyle\lim_{Q \to P} \left(\dfrac{QR}{PR}\right)$ = 8.

C.2 Gradient of a curve at any point

Now, to find the gradient at any point on a curve we will use exactly the same method as before. A point Q is taken on the curve whose x co-ordinate (or abscissa) is slightly greater than that of P.

Here we introduce a new notation, using a symbol to represent a small increase (or increment) in the variable. The small increase in the variable x is denoted by the symbol δx, called *delta x*.

Similarly, a small increment in y is denoted by δy, called *delta y*.

We let the small increase in x, PR, be δx and the corresponding small increase in y, QR, be δy.

The x coordinate of Q is $x + \delta x$ and its y co-ordinate is $y + \delta y$.

P lies on the curve $y = x^2$, and Q lies on the curve with y co-ordinate,

$$y + \delta y = (x + \delta x)^2 = x^2 + 2x\,\delta x + (\delta x)^2$$

So subtracting: $(y + \delta y) - \delta y$ yields QR, or

$$\delta y = 2x\delta x + (\delta x)^2$$

So the gradient of the chord PQ generally $= $ QR/PR $= \dfrac{\delta y}{\delta x} = \dfrac{2x\delta x + (\delta x)^2}{\delta x} = 2x + \delta x.$

Now as Q approaches P, δx will approach zero.

Thus the gradient of the chord PQ will become the instantaneous gradient of the curve at P.

Hence the gradient of the curve at P is the value to which the fraction $\dfrac{\delta y}{\delta x}$

approaches as δx approaches zero.

ie the gradient of the curve at $P = \dfrac{lim}{\delta x \to 0}\left(\dfrac{\delta y}{\delta x}\right)$

$= \dfrac{lim}{\delta x \to 0}(2x + \delta x) = 2x$

As $\delta x \to 0$ then $\dfrac{\delta y}{\delta x} \to \dfrac{dy}{dx}$, which is the gradient found by differentiation.

Hence if $y = x^2$ the gradient $\dfrac{dy}{dx} = 2x$

For the example given in C.1, the gradient $\dfrac{dy}{dx} = 2 \times 4 = 8$ which is in agreement with the specific value chosen, demonstrating this is a very general approach.

C.3 Differentiation from first principles

The previous general method of differentiation can be applied to obtain a gradient or differential coefficient for any equation, as long as we proceed in exactly the same way as we did to obtain the gradient of the curve in C.2.

E.g. if $y = x^2 + x + 4$, find the gradient $\dfrac{dy}{dx}$ from first principles.

Let x increase by a small amount δx and let the corresponding increase in y be δy.

Then as values of x and y are connected by the relation:

$y = x^2 + x + 4$

$y + \delta y = (x + \delta x)^2 + (x + \delta x) + 4$

And by subtraction as shown previously: $\delta y = 2x\delta x + \delta x^2 + \delta x$

$$= (2x + 1)\delta x + \delta x^2$$

Dividing by δx

$$\frac{\delta y}{\delta x} = (2x + 1) + \delta x$$

In order to obtain $\dfrac{dy}{dx}$ let δx tend to zero

Then $\dfrac{\delta y}{\delta x} \to \dfrac{dy}{dx}$ and $\dfrac{dy}{dx} = \dfrac{lim}{\delta x \to 0}(2x + 1) + \delta x$

Therefore $= \dfrac{dy}{dx} = 2x + 1$

The slope or gradient can be used to find both maxima and minima associated with an equation, as at the points of maxima/minima the gradient will be zero.

Hence $\dfrac{dy}{dx} = 0$ for maxima or minima. For example, for equation $y = x^2 + x + 4$, if the gradient is $2x + 1$ for maxima or minima:

$2x + 1 = 0$ so:

$2x = -1$

And $x = -0.5$

The maxima or minima will have the value $y = (-0.5)^2 - 0.5 + 5 = 4.75$.

Now the question arises: is it a maxima or a minima? Of course, changing the value of x a little about this point will quickly demonstrate that it is in fact a minima but there is an easier way if we consider the gradient or slope of the gradient itself.

Returning to figure C.1, if we consider point P to be at the minimum of the curve, the gradient will be zero, but the instantaneous gradient of the change as we move to the right will be to increasingly larger values and so will be positive. Further discussion of differential calculus (differentiation) is to be found in chapter 11 of Reeds Marine Engineering and Technology Series, Volume 1: *Mathematics for Marine Engineers* (ISBN 978–1-4081–7555–2).

Hence if $y = -x^2 - x + 4$

$$\frac{dy}{dx} = -2x - 1$$

And $\dfrac{d^2y}{dx^2} = -2$, which is negative. So the point where

$-2x - 1 = 0$

Thus $x = -0.5$ will actually be a maximum. It is left to the reader to show this by substitution of actual values.

Appendix D

Natural Logarithms and the Exponential Function

The function $f(x) = e^x$ is called the exponential function and its inverse is the natural logarithm, or logarithm to base e. The natural logarithm of a positive number k can also be defined as the area under the curve $y = 1/x$ between $x = 1$ and $x = k$, in which case, e is the number whose natural logarithm is 1 (figure D.1).

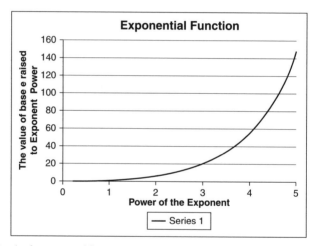

Figure D.1: *Graph of exponential function.*

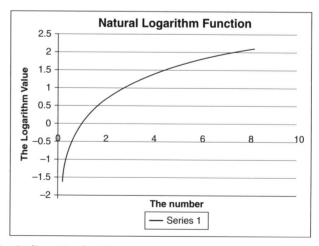

Figure D.2: *Graph of logarithm function.*

Figure D.2 shows a graph of the natural logarithm function. The function slowly approaches infinity as x increases and quickly falls towards negative infinity as x approaches zero asymptotically, as shown.

The *natural logarithm* of a number is its logarithm to base e, where e is an irrational and transcendental constant roughly equal to 2.7182 (4 decimal places). The natural logarithm of x is generally written as: ln x or $\log_e x$. The natural logarithm of x is the power to which e must be raised to equal x. For example, ln (4.3) is 1.458615, because $e^{1.458615...} = 4.3$. The natural log of e itself, ln(e), is 1, because $e^1 = e$, while the natural logarithm of 1, ln(1) = 0, since $e^0 = 1$ (ln $e^0 = 0 \times$ ln $e^0 = 0$).

Natural logarithms are used to solve equations in which the unknown quantity appears as the exponent of some other quantity. For example, logarithms are often used to find the half-life decay constant common in radioactive problems, or other exponential decay/growth problems in capacitor and inductor circuits. They are important in many branches of mathematics and the sciences and are used in finance to solve problems involving compound interest. The number e is an important mathematical constant that is the base of the natural logarithm. It is approximately equal to 2.71828 and is the limit of the expression: $(1 + 1/n)^n$ as n approaches infinity, an expression arising in the study of compound interest, and can be calculated as the sum of the infinite series:

$$e = \sum_{n=0}^{\infty} \frac{1}{n!} = 1 + \frac{1}{1} + \frac{1}{1 \cdot 2} + \frac{1}{1 \cdot 2 \cdot 3} + \ldots$$

Sometimes e is called *Euler's number* or *Euler's constant* after the Swiss mathematician Leonhard Euler (1707–1783). The number e is also known as *Napier's constant* and is of great mathematical importance alongside: 0, 1, π, and i. All five of these numbers play important and recurring roles across mathematics, and are the five constants appearing in one formulation of *Euler's identity*: $e^{i\pi} + 1 = 0$, which is a truly remarkable equation when you begin to think about all the mathematics contained within it.

Like the constant π, e is irrational: it is not just a ratio of integers and it is transcendental (without end). It is also not the root of *any* non-zero polynomial (having an equation form e.g. $x^2 + x^4$) with rational coefficients. The numerical value of e truncated to 15 decimal places is: 2.718281828459045.

D.1 History and accuracy of the natural logarithm

The concept of the natural logarithm was worked out by Gregoire de Saint-Vincent and Alphonse Antonio de Sarasa prior to 1649. Nicholas Mercator (famous for maps and charts) made mention of the natural logarithm in his work *Logarithmotechnia*, published in 1668.

The graph of the natural logarithm function shown earlier enables us to determine some of the basic characteristics that logarithms to any base have in common. Primarily, the logarithm of the number one is zero, and the logarithm as we approach zero approaches minus infinity. What makes natural logarithms unique is that there is a single point where all logarithms are zero, namely the logarithm of the number 1. At that specific point, the slope of the curve of the graph of the natural logarithm is exactly 1. Methods for computing the value of e can be thought of as resulting from the pursuit of this condition. One way of visualising this is to understand that, for any numerical value close to 1, the natural logarithm can be computed easily by subtracting 1 from the numerical value. For example, the natural logarithm of 1.01 is 0.01 to an accuracy better than 5 parts per thousand. With similar accuracy, one can state that the natural logarithm of 0.99 is -0.01.

The accuracy of this concept increases as one approaches 1 more closely, and is completely accurate at 1. To the same extent that the number 1 itself is a number common or universal to all counting systems, so also the natural logarithm is independent of all counting systems. In the English language, the term adopted to encapsulate this concept is the word *natural*. $\log_e$ is a natural logarithm because it arises so often in mathematics.

It might seem that the likeliest human numbering system would use base 10, and would be more obvious than base e. But mathematically, the number 10 is not very significant. Its use culturally, as the basis for many societies' numbering systems, arises from the typical number of human fingers. Other cultures have based their counting systems on numbering choices such as 5, 8, 12, 20, and of course the Babylonian counting system, upon which our own system relies heavily, uses 60 (60 seconds in a minute, 60 minutes in an hour, 360 degrees in a circle, etc.).

Solutions To Numerical Questions By Chapter

Chapter 1

2.

Using $V = f\lambda$ and rearranging the equation for the wavelength.

$\lambda = V/f$ and substituting for the values given.

    a.   so $\lambda = 3 \times 10^8/(9 \times 10^9) = 0.033$ m

    b.   so $\lambda = 3 \times 10^8/(85 \times 10^9) = 3.529$ m

    c.   so $\lambda = 3 \times 10^8/(1 \times 10^{12}) = 3 \times 10^{-4}$ m

3. Using the equivalence relationship:

$$\frac{t}{T} = \frac{\phi}{360}$$

So $t = \dfrac{\phi}{360} \times T = \dfrac{45}{360} \times T = 0.125\,T$ seconds

4. Ratio of the speed of light to the speed of sound in water approximately is given by: ratio $= \dfrac{c}{1500} = \dfrac{3 \times 10^8}{1500} = 2 \times 10^5$

5. Consider speed $v = \dfrac{\text{distanced}}{\text{time t taken to cover this distance}}$, given that the sound and light are radiated simultaneously, with the light travelling faster and thereby arriving first. This starts the clock of reception. The sound of thunder arrives 8 seconds later. The distance travelled in both cases is identical distance x.

Thus if light takes t_1 to arrive and the sound takes t_2 to arrive, the difference in time (8 seconds) will be equal to $t_2 - t_1$

Now from our speed equation above $t_1 = \dfrac{x}{c}$ and $t_2 = \dfrac{x}{v}$

Thus $8 = t_2 - t_1 = \dfrac{x}{v} - \dfrac{x}{c}$ or $8 = x\left(\dfrac{1}{v} - \dfrac{1}{c}\right)$. Now, if we ignore the second term as being

significantly less than the first term, so $8 = x\left(\dfrac{1}{v}\right)$ and hence:

$x = 8v = 8 \times 1500 = 1200$ m *or* 12 km

6. Using $V = f\lambda$ and rearranging the equation for the frequency:

$f = V/\lambda = \dfrac{1520}{0.03} = 50666.7$ Hz

7. Using $f = \dfrac{1}{T}$ and rearranging for the period T:

Using $T = \dfrac{1}{f} = \dfrac{1}{60} = 0.017$ seconds

8. $V = f\lambda$ so $\dfrac{dV}{dt} = f \times \dfrac{d\lambda}{dt} + \dfrac{df}{dt} \times \lambda$ if both variables are changing with respect to time.

9. $V_p = \dfrac{d\omega}{dk}v = \dfrac{\omega}{k}$ so: $\omega = vk$

$\dfrac{d\omega}{dk} = \dfrac{d(vk)}{dk} = \dfrac{vdk}{dk} + \dfrac{kdv}{dk}$ so $\dfrac{d\omega}{dk} = v + \dfrac{kdv}{dk}$

If $V_g = \dfrac{\omega}{k} = v$ and $V_p = \dfrac{d\omega}{dk}$ Then we can rewrite the previous

equation as: $V_p = V_g + \dfrac{kdv}{dk}$

10. Since $V_g = V_0 e^{-kx-\omega t}$ and using $V_p = V_g + \dfrac{kd\,v}{d\,k}$

Then: $V_p = V_0 e^{-kx-\omega t} + \dfrac{kd(V_0 e^{-kx-\omega t})}{dk} = V_0 e^{-kx-\omega t} + k(-k)V_0 e^{-kx-\omega t} = V_0 e^{-kx-\omega t}(1 - k^2)$

Hence as k tends to zero, the group velocity and the phase velocity will tend to the same value.

Chapter 2

1. $y(t) = A\cos(\omega t - \phi)$ differentiating with respect to time t

$$v(t) = \frac{dy(t)}{dt} = -A\sin(\omega t - \phi)$$

$$a(t) = \frac{dv(t)}{dt} = -A\omega^2\cos(\omega t - \phi) \text{ Substituting for: } A\cos(\omega t - \phi)$$

Thus: $a(t) = -\omega^2 y(t)$

2. $y(t) = A\sin(\omega t)$

$$\frac{dy(t)}{dt} = A\omega\cos(\omega t) = 0 \text{ for maximum or minimum.}$$

Thus either $\omega = 0$ as a trivial solution or $\cos(\omega t) = 0$ from which:
$\omega t = \cos^{-1}(0) = 90$ degrees.

To find whether this is a maximum or minimum, we need to look at the second
derivative, or rather: $a(t) = \dfrac{d^2y}{dt^2}$

Now: $a(t) = \dfrac{d^2y(t)}{dt^2} = -A\omega^2\sin(\omega t)$ for a maximum or minimum to occur and will be
+ve for minimum and −ve for maximum.

In the case above, $a(t) = \dfrac{d^2y(t)}{dt^2} = -A\sin(90) = -A\omega^2$, which is negative, making this
a maximum, which of course can be proven by inspection of the sine function.

3. $y(t) = A\cos(\omega t) + B\sin(\omega t)$

$$v(t) = \frac{dy(t)}{dt} = -A\omega\sin(\omega t) + B\omega\cos(\omega t) \text{ and,}$$

$$a(t) = \frac{d^2y(t)}{dt^2} = -A\omega^2\cos(\omega t) - B\omega^2\sin(\omega t)$$

So:

$3 = -A \omega \sin(\omega t) + B\omega \cos(\omega t)$, (1) and

$6 = -A \omega^2\cos(\omega t) - B \omega^2\sin(\omega t)$ (2)

From equation (1)

$$A \omega \sin(\omega t) = B\omega \cos(\omega t) - 3$$

Hence: $\sin(\omega t) = \dfrac{B\omega \cos(\omega t) - 3}{A \omega}$

And substituting into (2) for $\sin(\omega t)$

$$6 = -A \omega^2\cos(\omega t) - B \omega^2\frac{(B \omega \cos(\omega t) - 3)}{A \omega}$$

And simplifying:

$$6 = -A \omega^2\cos(\omega t) - B \omega\frac{(B \omega \cos(\omega t) - 3)}{A}$$

$$6 = -A \omega^2\cos(\omega t) + \frac{3B \omega}{A} - \frac{B^2\omega^2\cos(\omega t)}{A}$$

$$\frac{B^2\omega^2\cos(\omega t)}{A} + A \omega^2\cos(\omega t) - \frac{3B \omega}{A} + 6 = 0$$

Which simplifies further to:

$$\cos(\omega t)\left(\frac{B^2\omega^2}{A} + A \omega^2\right) - \frac{3B \omega}{A} + 6 = 0$$

or $\cos(\omega t) = \dfrac{+\left(\frac{3B \omega}{A} - 6\right)}{\left(\frac{B^2\omega^2}{A} + A \omega^2\right)}$

so: $\omega t = \cos^{-1}\left(\dfrac{+(\frac{3B\,\omega}{A} - 6)}{(\frac{B^2\omega^2}{A} + A\,\omega^2)}\right)$

and thus $t = \dfrac{1}{\omega}\cos^{-1}\left(\dfrac{+(\frac{3B\,\omega}{A} - 6)}{(\frac{B^2\omega^2}{A} + A\,\omega^2)}\right)$ providing a general solution for all possible

values of A and B for a given ω.

4. For a pendulum $\omega = \sqrt{\dfrac{g}{l}}$ $\omega = 2\pi f$ and $\omega = \sqrt{\dfrac{g}{l}} = 2\pi \times 2$

Thus $\dfrac{g}{l} = (4\pi)^2$ and so: $l = \dfrac{g}{(4\pi)^2} = 0.06\ m$

5. $c = \sqrt{\dfrac{g\lambda}{2\pi}} = \sqrt{\dfrac{9.81 \times 50}{2\pi}} = 8.84$ metres per second.

6. In shallow water: $c = \sqrt{gz} = \sqrt{9.81 \times 10} = 9.9$ metres per second. Note z dependence only, not λ!

7. $c_d = \sqrt{\dfrac{g\lambda}{2\pi}}$ and $c_s = \sqrt{gz}$ so the ratio is given by: $\dfrac{c_d}{c_s} = \dfrac{\sqrt{\frac{g\lambda}{2\pi}}}{\sqrt{gz}} = \sqrt{\dfrac{g\lambda}{2\pi gz}} = \sqrt{\dfrac{\lambda}{2\pi z}}$

8. $A = \dfrac{\lambda}{2\pi}\exp\left(-\dfrac{2\pi z}{\lambda}\right)$ substituting for the values

given: $A = \dfrac{8}{2\pi}\exp\left(-\dfrac{2\pi 4}{8}\right) = \dfrac{4}{\pi}\exp(-\pi) = 0.055\ m$

9. $A = \dfrac{\lambda}{2\pi}\exp\left(-\dfrac{2\pi z}{\lambda}\right)$

The gradient of the curve is given by: $\dfrac{dA}{dz}$ so:

$$\frac{dA}{dz} = -\frac{2\pi\lambda}{2\pi\lambda}\exp\left(-\frac{2\pi z}{\lambda}\right) = -\frac{2\pi}{\lambda}A = -kA$$

10.

Since: $\dfrac{p_{Initial}}{\rho} + \dfrac{v_{Initial2}}{2} + gH_{Initial} = \dfrac{p_{Final}}{\rho} + \dfrac{v_{Final2}}{2} + gH_{Final}$

Then: $\dfrac{p_{Initial}}{\rho} + \dfrac{v_{Initial2}}{2} + gH_{Initial} = \dfrac{p_{Final}}{\rho} + \dfrac{4v_{Initial2}}{2} + g\dfrac{H_{Initial}}{2}$

$\dfrac{p_{Final}}{\rho} + gH_{Initial} - g\dfrac{H_{Initial}}{2} - 2v_{Initial2} + \dfrac{v_{Initial2}}{2} + \dfrac{p_{Initial}}{\rho}]$

Rearranging for final pressure:

$$p_{Final} = \rho\left(g\frac{H_{Initial}}{2} - \frac{3v^2}{2} + \frac{p_{Initial}}{\rho}\right)$$

Chapter 3

1. (4) 10^{10} GHz SHF, (6) 3×10^{18} X-rays, (1) 15 kHz *echo sounder waves*, (3) 3×10^9 radar, (2) 2×10^5 radio waves, and (5) 3×10^{15} Hz blue light.

2. The following list shows five different wave motions:

3 GHz radar $\lambda = \dfrac{3\times10^8}{3\times10^9} = 0.1\ m$

Ultraviolet light typically 200–400 nm

3 MHz radio signal $\lambda = \dfrac{3\times10^8}{3\times10^6} = 100\ m$

15 kHz sonar, and $\lambda = \dfrac{1500}{15\times10^3} = 0.1\ m$

1500 Hz sound in air $\lambda = \dfrac{330}{1500} = 0.22\ m$

c. Both the 3 GHz radar and the 15 kHz sonar transmission have the shortest wavelength.

d. Ultraviolet radiation has a higher frequency than visible light.

3. Placed in order of frequency, starting with the lowest:

(6) 5 GHz radar

(5) 3 micron infra-red radiation $f = \dfrac{3 \times 10^8}{3 \times 10^{-6}} = 1 \times 10^{14} Hz$

(2) 20 kHz sonar

(1) 4 kHz sound in air

(4) blue light $= \dfrac{3 \times 10^8}{400 \times 10^{-6}} = 7.5 \times 10^{11} Hz$ and

(2) 2000 m radio $= \dfrac{3 \times 10^8}{2000} = 150 \ kHz.$

4. $f = \dfrac{3 \times 10^8}{600 \times 10^{-9}} = 5 \times 10^{14} Hz$

and $E = hf = 6.62 \times 10^{-34} \times 5 \times 10^{14} = 3.31 \times 10^{-19} J$

5. If $1 \, eV = 1.6 \times 10^{-19} J$

So: $Energy \ in \ eV = \dfrac{3.31 \times 10^{-19}}{1.6 \times 10^{-19}} = 2.069 eV$

6. Using the equation: $I_x = I_o e^{-ax}$ and rearranging:

$\dfrac{I_x}{I_o} = e^{-ax}$

$\dfrac{1}{7} = e^{-ax}$

Taking inverse logarithms:

$-\dfrac{\ln(\frac{1}{7})}{a} = x$

So for $a = 0.76\ m^{-1}$

$$x = -\frac{\ln(\frac{1}{4})}{0.76} = 1.824\ m$$

7. Using the equation $I_x = I_o e^{-ax}$ and differentiating with respect to distance x:

$$\frac{dI_x}{dx} = (-a)I_o e^{-ax}\ \text{Hence:}$$

$$\frac{dI_x}{dx} = -aI_x$$

8. Using:

$$\frac{dI_x}{dx} = -aI_x$$

$$\frac{1}{I_x}\frac{dI_x}{dx} = -a$$

So:

$$a = -\frac{1}{I_x}\frac{dI_x}{dx}$$

and substituting for values given:

$$a = -\frac{1}{12.4}3 = 0.24\ m^{-1}$$

9. Energy in an individual X-ray photon
$E = hf = 6.62 \times 10^{-34} \times 2.4 \times 10^{19} = 1.5888\ 10^{-14} J$

$$\text{Total number of X-ray photons} = \frac{Total\ Energy}{Energy\ in\ an\ individual\ X-ray\ photon} =$$

$$\frac{1.6\ \times\ 10^{-11}}{1.5888\ 10^{-14}} = 1007\ \text{X-ray photons.}$$

Chapter 4

2. Using $I = I_0{}^2\cos^2\theta \dfrac{dI}{d\theta} = -I_0{}^2 2\ \cos\theta\sin\theta = -I_0{}^2\sin2\theta$

If $\cos\theta = 0$, one solution is $\theta = 90$ degrees

or if $\sin\theta = 0$, another solution is $\theta = 0$ degrees

$$\frac{d^2I}{d\theta^2} = -2I_0{}^2\cos2\theta$$

So: taking a solution $\theta = 90$ degrees, $\dfrac{d^2I}{d\theta^2} = -2I_0{}^2\cos180 = 2I_0{}^2$, which is positive and

is therefore a minimum, i.e. the polariser and the analyser are crossed at 90 degrees to each other.

And also taking a solution $\theta = 0$ degrees, $\dfrac{d^2I}{d\theta^2} = -2I_0{}^2\cos0 = -2I_0{}^2$ which is negative

and is therefore a maximum, i.e. the polariser and analyser are aligned in the same direction.

6. $N_i \sin\theta_i = N_t \sin\theta_t$

$N_1 \sin\theta_c = N_2 \sin 90$

$$\sin\theta\ c = \frac{N_2}{N_1} = \frac{1.0003}{2.4} = 0.41679$$

So $\theta\ c = \sin^{-1}\left(\dfrac{N_2}{N_1}\right) = \sin^{-1}0.41679 = 24.63$ degrees

7.

i) Using $\lambda = dx/L$ and substituting:

$\lambda = 2.5 \times 10^{-3} \times 0.05 \times 10^{-2} /2.3 = 5.4348 \times 10^{-7} = 543$ nm

ii) Rearranging $L = dx/\lambda$ and substituting:

$= 2 \times 10^{-3} \times 2 \times 10^{-2} /400 \times 10^{-9} = 5$ m

8. Using: $N_i \sin \theta_i = N_t \sin \theta_t$

$1.0003 \sin 30 = 1.333 \sin \theta_t$

Thus $\sin \theta_t = 0.375$ and $\theta_t = 22.037$ degrees

9.

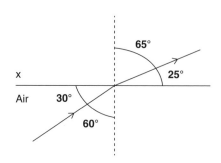

Using: $N_i \sin \theta_i = N_t \sin \theta_t$

$$N_t = \frac{Ni \, \sin\theta \, i}{\sin\theta \, t}$$

Substituting $N_t = \dfrac{1.0003 \sin 60}{\sin 65} = 0.9558.$

Note: This is a value less than 1!

10. Using: $N_i \sin \theta_i = N_t \sin \theta_t$
$1.333 \sin 20 = 1.52 \sin \theta_t$
Thus $\sin \theta_t = 0.2999$ and $\theta_t = 17.454$ degrees

Chapter 5

2. Using the equation: $I = \dfrac{P}{4\pi R^2}$

Substituting: $I = \dfrac{25}{4\pi \, 60^2} = 5.53 \times 10^{-4} Wm^{-2}$

3.

$$I_1 = \frac{P_1}{4\pi R_1^2} \ (3.1) \text{ and } I_2 = \frac{P_2}{4\pi R_2^2} \ (3.2)$$

Dividing equation 3.1 by equation 3.2 gives:

$$\frac{I_2}{I_1} = \frac{P_2}{P_1} \times \frac{R_1^2}{R_2^2} \quad \textbf{(5.3)}$$

Substituting for the values given:

$$\frac{I_2}{I_1} = \frac{1.3\,P}{P} \times \frac{75\ km^2}{55\ km^2}$$

$$I_2 = \frac{1.3\,I_1 P}{P} \times \frac{75\ km^2}{55\ km^2}$$

$$I_2 = \frac{1.31 \times 4 \times 10^{-2} P}{P} \times \frac{75\ km^2}{55\ km^2}$$

$$I_2 = 6.24 \times 10^{-2} \times \frac{75^2}{55^2} = 0.116\ W\,m^{-2}$$

5. For echoes:

$$\frac{I_2}{I_1} = \frac{P_2}{P_1} \times \frac{R_1^4}{R_2^4}$$

$$\frac{I_2}{I_1} = \frac{P}{P} \times \frac{50^4}{150^4} = \frac{50^4}{150^4} = 0.01234$$

6.

For echoes: $\frac{I_2}{I_1} = \frac{P_2}{P_1} \times \frac{R_1^4}{R_2^4}$ At the maximum detection range, the receiver will receive the minimum intensity I_{min} for that receiver.

$$\frac{I_{min}}{I_{min}} = \frac{1.25P}{P} \times \frac{110^4}{R_2^4}$$

$$1 = 1.25 \times \frac{110^4}{R_2^4}$$

$R_2^4 = 1.25 \times 110^4$

$R_2 = 1.25 \times 110^4 = 116.3$ km

7. $I_s \propto \dfrac{1}{\lambda^4}$ Then: $I_{450nm} \propto \dfrac{1}{450^4}$ and $I_{650nm} \propto \dfrac{1}{650^4}$

Dividing these two inequalities:

$$\dfrac{I_{450nm}}{I_{650nm}} = \dfrac{650^4}{450^4} = 1.444^4 = 4.353$$

8. Using the equation: $C(\lambda) = A(\lambda) + B(\lambda)$

$A(\lambda) = 0.094$

$B(\lambda) = 0.002$

$C(\lambda) = 0.094 + 0.002 = 0.096$

$I_{out} = I_{in}\, e^{-c(\lambda)x} = 25 \times e^{-0.096 \times 20} = 3.67$ Wm^{-2}

9.

$I \propto \dfrac{1}{\lambda^4}$ and also $I \propto \dfrac{1}{d^2}$ combining both relationships:

$I \propto \dfrac{1}{\lambda^4} \times \dfrac{1}{d^2}$

$I = k\dfrac{1}{\lambda^4} \times \dfrac{1}{d^2}$

Differentiation with respect to both wavelength and scatterer size d respectively:

$\dfrac{dI}{d\lambda} = -4k\dfrac{1}{\lambda^5} \times \dfrac{1}{d^2}$ and $\dfrac{dI}{dd} = -2k\dfrac{1}{\lambda^4} \times \dfrac{1}{d^3}$

when $dI/dd = dI/d\lambda - 4k\dfrac{1}{\lambda^5} \times \dfrac{1}{d^2} = -2k\dfrac{1}{\lambda^4} \times \dfrac{1}{d^3}$

Thus: $\dfrac{2}{\lambda} = \dfrac{1}{d}$ Hence $\lambda = 2d$

10. Using: $I_{out} = I_{in} e^{-C(\lambda)x}$

$C(\lambda) = A(\lambda) + B(\lambda) = 0.053 + 0.002 = 0.055$

So: $I_{out} = 0.75 \times e^{-0.055 \times 13} = 0.37 \, Wm^{-2}$

Chapter 6

2. $D_{horizontal} = \dfrac{60}{a_{horizontal}} = \dfrac{60c}{fa_{horizontal}} = \dfrac{60 \times 3 \times 10^8}{9.5 \times 10^9 \times 1} = 1.89 \, m$

   $D_{vertical} = \dfrac{60}{a_{vertical}} = \dfrac{60c}{f \times a_{vertical}} = \dfrac{60 \times 3 \times 10^8}{9.5 \times 10^9 \times 20} = 0.095 \, m$

3. $a_{horizontal} = \dfrac{60}{D_{horizontal}} = \dfrac{60c}{f \times D_{horizontal}} = \dfrac{60 \times 3 \times 10^8}{9.5 \times 10^9 \times 20} = 2.42°$

   $a_{vertical} = \dfrac{60}{D_{vertical}} = \dfrac{60c}{f \times D_{vertical}} = \dfrac{60 \times 3 \times 10^8}{3.1 \times 10^9 \times 0.20} = 23.23°$

6. No. of lines per metre $= 6 \times 10^5$

Spacing $d = \dfrac{1}{6 \times 10^5} = 1.67 \times 10^{-6} \, m$

For the *nth* constructive order for a grating

$d \sin\theta = n\lambda$

The maximum value of θ incident is 90°, so sin (90°) = 1

Taking the spectral range for white light to be between 400 nm and 700 nm, the midpoint of the visible spectrum observed will be about 550 nm or 550×10^{-5} m.

So the *maximum* possible value of n is:

$$n = \frac{d\sin\theta}{\lambda} = \frac{1.67 \times 10^{-6} \times \sin 90}{550 \times 10^{-9}} = 3.04 \text{ (2 decimal places)}$$

Therefore three complete orders should be seen.

8. The angular separation $= 60\lambda / D = 60 \times \dfrac{2}{1 \times 10^3} = 0.12$ degrees.

9. The array angular separation $= 60\lambda/D = 60\lambda/(n-1)d =$

$$60 \times \frac{2}{(40-1) \times 1 \times 10^3} = 3.08 \times 10^{-3} \text{degrees.}$$

10. Using the equation: the array angular separation $0.001^0 = 60\lambda/(n-1)d =$

$$60 \times \frac{0.05}{(40-1) \times d}$$

Rearranging for d: $d = 60 \times \dfrac{0.05}{(40-1) \times 0.001} = 76.92$ m apart.

Chapter 7

4. See the diagram following:

$$dB - dA = (2n-1)\frac{\lambda}{2} \text{ for minimum where } n = \pm 1, \pm 2, \ldots, \pm m$$

When $n = 2$, $dB - dA = 3\dfrac{\lambda}{2}$

$dB - dA =$ extra path difference $= (x + dx) - (x - dx) = 2dx$

Thus $2dx = 3\dfrac{\lambda}{2}$ so $dx = 3\dfrac{\lambda}{4}dx$

6. Using the representation: $\dfrac{t}{T} = \dfrac{d}{\lambda} = \dfrac{\phi}{360}$

For the path difference: $\dfrac{t}{T} = \dfrac{d}{\lambda}$ so t= $\dfrac{dT}{\lambda}$ and

The equivalent time delay = $\dfrac{\lambda / 3T}{\lambda}$ = T/3 and

for the phase difference $\dfrac{t}{T} = \dfrac{\phi}{360}$ so t= $\dfrac{\phi T}{360} = \dfrac{180T}{360}$ = T / 2

and considering all the time delays, path differences and phase differences in terms of equivalent time, then:

The total time delay = T/4 + T/3 +T/2 = 11T/4

7. The time delay equivalence can be found from the expression: $\dfrac{t}{T} = \dfrac{d}{\lambda}$. The time delay required if the path difference between two adjacent sources is 2 cm will be t= $\dfrac{dT}{\lambda} = \dfrac{0.02\ T}{0.10}$ = T / 5.

8. The total number of sources is the square of the number of sources along any row or column, ie the number of sources n in any row or column is related to the total number of sources N in a square array by the equation: $N = n^2$

So $n = \sqrt{N}$

Hence $n = \sqrt{900}$ = 30

Now, the number of *spaces* is related to the number of sources by the equation:

number of spaces = number of sources – 1

So the number of spaces = 30–1 = 29

Each space d has a value of $\dfrac{\lambda}{2} = \dfrac{3}{2}$ = 1.5 cm

The beam width equation gives: $a = \dfrac{60\lambda}{D} = \dfrac{60\lambda}{(n-1) \times d} = \dfrac{60\lambda}{29 \times \frac{\lambda}{2}} = \dfrac{60}{99} = 4.138$

degrees.

9. Given that speed = path difference/ time delay $c = \dfrac{path\ difference}{time\ delay}$

Therefore: $time\ delay = \dfrac{0.02}{3 \times 10^8} = 6.67 \times 10^{-11}$

Chapter 8

3. Since: $\Delta f = \dfrac{V_{relative}}{c} f_{transmitted}$

$\Delta f = \dfrac{10}{3 \times 10^8} \times 10 \times 10^9 = 333.3\ metres\ per\ second$

4. For echoes: $\Delta f\ echoes = \dfrac{2V_{relative}}{c} f_{transmitted}$

So rearranging: $V_{relative} = \dfrac{c \times f\ echoes}{2f_{transmitted}}$

And substituting: $V_{relative} = \dfrac{3 \times 10^8 \times 3 \times 10^3}{2 \times 9.54 \times 10^9} = 47.17$ metres per second.

5. Use the following expression for relative velocity:

$V_{relative} = V_{source}\cos\theta_{source} - V_{target}\cos\theta_{target}$

$V_{relative} = 10 \times \cos30 - 25 \times \cos120 = 21.16\ metres\ per\ second$

And using $\Delta f\ echoes = \dfrac{2V_{relative}}{c} f_{transmitted}$: and substituting:

$\Delta f\ echoes = \dfrac{2 \times 21.16}{3 \times 10^8} \times 3.5 \times 10^9 = 493.73\ Hz$

6. Using the power ratio expression in decibels $= 10 \log (P_2 / P_1)$

So: $+ 35 = 10 \log (P_2 / 0.038)$ and taking antilogs $P_2 = 0.038 \times 10^{3.5}$

$= 120.17 \, W$

7. So the ratio in decibels $= 10 \log (P_2 / P_1)$

Substituting for the figures given:

decibels $= 10 \log (100/5) = 13.01$

Chapter 9

1. $(a + b)^3 = (a + b) \times (a + b) \times (a + b) = (a^2 + ab + ab + b^2) \times (a + b) =$

$a^3 + a^2b + a^2b + a b^2 + a^2b + a b^2 + ab^2 + b^3 = a^3 + 3a^2b + 3a b^2 + b^3$

Note the symmetry in the terms of the final answer.

2. Expand $(a + b)^2 \times (c + d) = (a^2 + 2ab + b^2) \times (c + d) = a^2c + a^2d + 2abc + 2abd + b^2c + b^2d$

3. $a^{-n} \times a^m = 2^{-3} \times 2^4 = 2$

4. $a^{-n} \times a^m = 3^{-4} \times 3^{-2} = 3^{-6} = 1.3717 \times 10^{-3}$

5. $a^n \times a^m = 1.6^{2.4} \times 1.6^{3.3} = 1.6^{5.7} = 14.57$

6. $y = mx + c$ so

$y - c = mx$ and thus $x = \dfrac{y - c}{m}$

8.

(i)   3.19478

(ii)   1.37402

(iii)   1.324.46

9. Find the gradient (dy/dx) for $y = (a + x)^2$

$(dy/dx) = 2(a + x)$

10. To find the derivative of $f(x) = \ln(3x - 4)$, use the chain rule.

We have $(3x - 4)' = 3$ and $(\ln u)' = 1/u$. Putting this together gives:

$f'(x) = (3)(1/u)$

$$= \frac{3}{3x - 4}$$

Wave Motion Glossary

Amplitude (A) The maximum displacement of a particle from its undisturbed equilibrium position.

Anti-phase Two waves of equal frequency are said to be in *anti-phase* if they are half a cycle (180°, half a wavelength, or half a period) out of phase.

Beam width (α) The angular spread of a beam of electromagnetic radiation measured between the half power points (the 50% level or -3 dB points).

Critical angle (θ_c) The incident angle at a boundary between two different media, which gives an angle of refraction = 90°. The critical angle marks the boundary between refraction and Total Internal Reflection (TIR), when waves move from a low velocity medium to a higher velocity medium.

Cycle This is the basic repeating unit of a periodic wave, which consists of one complete oscillation or vibration of the wave.

Diffraction This is the phenomenon when waves bend around an object's edge or through a narrow slit. Diffraction causes spreading of the wave into the object's geometrical shadow or makes waves following the object's surface.

Electromagnetic wave A wave consisting of an oscillating electric field and oscillating magnetic field at right angles to each other and at right angles to the direction of propagation.

Electromagnetic spectrum The range of frequencies covered by all the electromagnetic waves.

Frequency (f) The number of wave cycles occurring in one second or the number of wave crests passing a point in one second. Frequency has units of Hertz (Hz).

Frequency domain diagram Also known as a frequency spectrum, this shows frequency content of a waveform on a graph of amplitude against frequency.

In phase Two waves of equal frequency are in phase when the phase difference between them is zero or a whole number of wave cycles.

Intensity (I) This is a measure of the rate of flow of energy per unit area of a surface at right angles to the direction of propagation of the wave. It is equal to the power of the source divided by the area over which this power is spread. Intensity has units of watts (W) per square metre (Wm^{-2}).

Interference A combination of two or more waves creates a composite waveform. The component waves can add together (constructive interference) or subtract and cancel (destructive interference).

Inverse square law A purely geometric effect showing intensity reduction by the spreading of electromagnetic waves with increasing distance from a source, the inverse square law means that intensity is proportional to $1/R^2$.

Inverse fourth power law For an echo, waves undergo the inverse fourth power law, spreading on both the outward and return journeys. Intensity varies inversely as the fourth power of the range; consequently, intensity of echoes will be proportional to $1/R^4$.

Longitudinal wave A wave in which oscillations of the medium through which the wave is travelling are along the direction of propagation, e.g. a sound wave travelling in air or a sonar underwater wave.

Monochromatic waves Electromagnetic waves of the same frequency. A source that produces waves of only one frequency is called a monochromatic source, e.g. a laser source is monochromatic.

Period, periodic time (T) This is the time taken to complete one wave cycle. The period is the reciprocal of the wave's frequency and is measured in units of seconds (s). $T = \dfrac{1}{f}$ where f is the wave frequency.

Phase Phase is the name given to how far through a wave cycle a particular point on a wave is. To have any meaning with a single waveform, the phase of a point must be compared to the phase of a fixed reference point, usually the origin. Commonly measured as the phase angle in degrees or radians.

Phased array A regular array usually in two dimensions of sources, which can be controlled together to provide a single beam of radiated energy and can also be used to detect any echoes that may arise.

Phase difference (φ) This is the difference in phase between two waves with the same frequency, using one of the waves as a reference.

Polarised wave An electromagnetic wave where there is a preferred orientation of the electric field. If all the electric fields are in one direction, the wave is said to be linearly or plane polarised.

Power (P) The rate at which a wave source performs work. Power is the rate of energy consumption. Power has units of watts (W).

Energy = Power × time

Progressive wave A moving wave that transfers energy or information from one point in space to another.

RADAR An echo-location method used by civilians and military forces to obtain vessel and 'target' ranges using electromagnetic waves, typically in the mm wave band. RADAR is an acronym for Radio Aid for Detection And Ranging.

Radian measure An alternative to degrees for measuring angles. There are 2π radians in a circle and so 2π radians = 360 degrees.

Reflection The return of a wave from a surface upon which it is incident. Direct reflection methods are used in radar and sonar echo location.

RefractionThis is the bending of a wave as it moves between different media. It occurs when the wave has a different velocity in each medium.

Sonar (SOund Navigation And Ranging) A similar technique to radar, which uses sound for echo location purposes.

Speed (V) The distance a wave front travels in unit time. Speed has units of metres per second (ms^{-1}). $V = f/\lambda$ for EM waves $c = f\lambda$. Velocity is a speed in a given direction.

Standing Wave A standing wave or stationary wave is a wave in a medium in which each point on the axis of the wave has an associated constant amplitude, due to the presence of both a forward travelling and backward travelling wave.

The locations at which the amplitude is minimum are called nodes, and the locations where the amplitude is maximum are called antinodes.

Superposition The principle whereby the resultant of two, or more, waves meeting at a point is found by summing the separate effects at that point for each wave on its own.

Time domain diagram This shows how a waveform changes with time on a graph of displacement against time.

Total internal reflection (TIR) When a wave is incident at the boundary between a low velocity medium and a higher velocity medium and its angle of incidence is greater than the critical angle, it will undergo total internal reflection.

Transverse wave A wave in which the direction of displacement is perpendicular to the direction of propagation, e.g. a mechanical wave such as a resonating ruler on the edge of a desk.

Unpolarised wave An electromagnetic wave composed of very many randomly oriented waves, which shows no preferred orientation of the electric fields.

Wavelength (λ) The distance between a point on a wave and the next identical point having the same phase. It is, therefore, the length of one cycle and is usually taken as the distance between successive crests or troughs. Wavelength has units of metres (m).

Index

Note: Page numbers followed by 'f' and 't' refer to figures and tables, respectively.